Homebrew Gaming and the Beginnings of Vernacular Digitality

Game Histories

edited by Henry Lowood and Raiford Guins

Debugging Game History: A Critical Lexicon, edited by Henry Lowood and Raiford Guins, 2016

Zones of Control: Perspectives on Wargaming, edited by Pat Harrigan and Matthew Kirschenbaum, 2016

Gaming the Iron Curtain: How Teenagers and Amateurs in Communist Czechoslovakia Claimed the Medium of Computer Games, Jaroslav Švelch, 2018

The Elusive Shift: How Role-Playing Games Forged Their Identity, Jon Peterson, 2020

Homebrew Gaming and the Beginnings of Vernacular Digitality, Melanie Swalwell, 2021

Homebrew Gaming and the Beginnings
of Vernacular Digitality

Melanie Swalwell

The MIT Press
Cambridge, Massachusetts
London, England

© 2021 Massachusetts Institute of Technology

All rights reserved. No part of this book may be reproduced in any form by any electronic or mechanical means (including photocopying, recording, or information storage and retrieval) without permission in writing from the publisher.

This book was set in Stone Serif and Stone Sans by Westchester Publishing Services. Printed and bound in the United States of America.

Library of Congress Cataloging-in-Publication Data is available.

Names: Swalwell, Melanie, 1972- author.
Title: Homebrew gaming and the beginnings of vernacular digitality / Melanie Swalwell, The MIT Press.
Description: Cambridge : The MIT Press, 2021. | Series: Game histories | Includes bibliographical references and index.
Identifiers: LCCN 2020037102 | ISBN 9780262044776 (Hardcover)
Subjects: LCSH: Video games--Australia--History. | Video games--New Zealand--History. | Video games--Design--History. | Computer games--Australia--History. | Computer games--New Zealand--History. | Computer games--Design--History. | New media art. | Homebrew (Video games)
Classification: LCC GV1469.3 .S955 2021 | DDC 794.80994--dc23
LC record available at https://lccn.loc.gov/2020037102

10 9 8 7 6 5 4 3 2 1

For Pearl

Our categories of knowledge are still too rustic and our analytic models too little elaborated to allow us to think the inventive proliferation of everyday practices. That is our regret. That there remains so much to understand about the innumerable ruses of the "obscure heroes" of the ephemeral, those walking in the city, inhabitants of neighbourhoods, readers and dreamers, the obscure Kitchen Women Nation, fills us with wonder.

—Michel de Certeau and Luce Giard (1983)

Contents

Series Foreword xi
Preface xiii

1 **Introduction** 1
2 **Discourses about Microcomputers** 23
3 **Micro Users as Makers** 49
4 **The Games** 81
5 **Hardware Hacking and Electronics** 129
6 **The Legacy of 1980s Homebrew** 147
7 **New Directions** 171

Notes 181
Homebrew Software Cited 195
Works Cited 199
Index 219

Series Foreword

What might histories of games tell us not only about the games themselves but also about the people who play and design them? We think that the most interesting answers to this question will have two characteristics. First, the authors of game histories who tell us the most about games will ask big questions. For example, how do gameplay and design change? In what ways is such change inflected by societal, cultural, and other factors? How do games change when they move from one cultural or historical context to another? These kinds of questions forge connections to other areas of game studies, as well as to history, cultural studies, and technology studies.

The second characteristic we seek in "game-changing" histories is a wide-ranging mix of qualities partially described by terms such as *diversity*, *inclusiveness*, and *irony*. Histories with these qualities deliver interplay of intentions, users, technologies, materials, places, and markets. Asking big questions and answering them in creative and astute ways strikes us as the best way to reach the goal of not an isolated, general history of games but rather of a body of game histories that will connect game studies to scholarship in a wide array of fields. The first step, of course, is producing those histories.

Game Histories is a series of books that we hope will provide a home—or maybe a launch pad—for the growing international research community whose interest in game history rightly exceeds the celebratory and descriptive. In a line, the aim of the series is to help actualize critical historical study of games. Books in this series will exhibit acute attention to historiography and historical methodologies, while the series as a whole will encompass the wide-ranging subject matter we consider crucial for the relevance of historical game studies. We envisage an active series with output that will reshape how electronic and other kinds of games are understood, taught, and researched, as well as broaden the appeal of games for allied fields such as history of computing, history of science and technology, design history, design culture, material culture studies, cultural and social history, media history, new media studies, and science and technology studies.

The Game Histories series will welcome but not be limited to contributions in the following areas:

- Multidisciplinary methodological and theoretical approaches to the historical study of games.
- Social and cultural histories of play, people, places, and institutions of gaming.
- Epochal and contextual studies of significant periods influential to and formative of games and game history.
- Historical biography of key actors instrumental in game design, development, technology, and industry.
- Games and legal history.
- Global political economy and the games industry (including indie games).
- Histories of technologies pertinent to the study of games.
- Histories of the intersections of games and other media, including such topics as game art, games and cinema, and games and literature.
- Game preservation, exhibition, and documentation, including the place of museums, libraries, and collectors in preparing game history.
- Material histories of game artifacts and ephemera.

Henry Lowood, Stanford University

Raiford Guins, Indiana University Bloomington

Preface

This book has had quite a gestation. Its origins lie in a consultancy I undertook for Te Manawa Museums Trust in New Zealand in 2004. Te Manawa was particularly interested in the local history of games in New Zealand. I had just arrived to take up my first full-time academic position and knew little about game history. Surely New Zealand's game history could not have been all that different from the rest of the world's, I reasoned. I turned out to be quite wrong, but it was the most productive mistake ever! Some of the fieldwork was undertaken during my time at Victoria University of Wellington, where the support I received both from the university budget and from my collaborators nurtured not only this project but also my development as a researcher. Working with Ian Welch and Susan Corbett, I got my first taste of how to make a multidisciplinary team really work.

My game history research first saw publication in interactive format for the scholarly interactive essay I developed during 2005–2006 for the online journal *Vectors: Culture and Technology in a Dynamic Vernacular* at the University of Southern California, and a couple of years later in the *Aotearoa Digital Arts* anthology, when I ventured to theorize what I called the art of amateur programming. I am grateful to Tara McPherson, Steve Anderson, and my collaborator Erik Loyer for the wonderful *Vectors* publishing opportunity, and Su Ballard and Stella Brennan for their appreciation that game history belonged in a book on digital arts.

Researching digital game history has made me into the type of patron who surely must stretch librarians' patience. I have had to make it my business to search for what was considered unlikely to be worth remembering and often has largely been forgotten. Traces of unofficial, ordinary culture do not always show themselves readily. That I have somehow still managed to find what I needed—in public libraries in Christchurch, Auckland,

and Wellington, the National Library of New Zealand / Alexander Turnbull Library (ATL), the State Library of New South Wales, and the Special Collections of Stanford University Libraries—is largely thanks to the guidance I have received from remarkable library staff. I am in their debt. I particularly appreciated the inspired suggestions of Barbara Brownlie early on, when I spent a morning 'discovering' things I didn't know I needed to find in the ATL's Ephemera collection. The firm belief Barbara had in me and this project provided early encouragement, and the junk mail catalogs that had been collected over the years proved to be so important. Given that many didn't see the point of opening game historical scholarship as an area, I remain grateful to all those—including Sydney Shep and Roy Shuker—who boosted and were enthusiastic about this program of research.

I have been the beneficiary of several awards that have directly contributed to realizing this book. Returning to Australia, I was honored to receive the 2009 Nancy Keesing Fellowship at the State Library of New South Wales, which allowed me to study their nonlending collection of 1980s code books, ironically never used in front of a microcomputer. That was the moment when figure (homebrew games for 1980s microcomputers) began to separate itself from ground (game history more generally). The award to the "Play It Again" research team of an Australian Research Council Linkage Project grant in 2011 was a high point, in conjunction with our museum partners the Australian Centre for the Moving Image, Ngā Taonga Sound & Vision, and the Berlin Computerspiele Museum (LP120100218). I was also the recipient of an ARC Future Fellowship to research "Creative Micro-computing in Australia, 1976–92," of which this homebrew book comprises but one part (FT130100391). Studying the wider history of creative microcomputing has enriched the current project, though also extending the timeline for its completion.

It has been an honor and a pleasure to present aspects of this research in several keynote addresses at the New Media Histories conference in Lodz, Poland, in 2014; the 2017 Digital Game Research Association conference in Melbourne; and the 2017 Central and Eastern European Game Studies conference in Trnava, Slovak Republic. I thank the organizers for their hospitality and thank the participants for their engagement.

I need to pay tribute to my interlocutors and informants. Some are not mentioned here but have contributed to my understanding in important ways. I thank the named informants for sharing their stories with me and

Preface

allowing me to include their accounts, which powerfully animate the archival materials. Some informants have become collaborators, and among them I acknowledge collectors Michael Davidson, Alan Laughton, Clinton Rowe, Andrew Stephen, and Aaron Wheeler, heroes of digital cultural heritage in New Zealand and Australia.

I have had the pleasure of working with some wonderful research assistants over the years, possessed of determination and great finding skills! Kudos to Janet Bayly, Nerina Bennett, Kerstin Grosch, Sarah McKenzie, James O'Connor, and Liz Larkin, project manager extraordinaire.

Colleagues from game history and beyond have generously shared their knowledge with me, listened to conference and seminar papers, read drafts, and helped me to refine my ideas. There are too many to name them all, but special thanks must go to Maria Garda, Graeme Kirkpatrick, Andreas Lange, David Murphy, and Jaroslav Švelch. The series editors, Henry Lowood and Raiford Guins, have been very supportive not just of this monograph but my wider project over many years. Noah Springer has been a thoughtful and responsive editor, guiding this book through to completion. I also thank the three anonymous reviewers for their comments and suggestions.

Finally, I'm fortunate to have some very special people in my life: collaborators and staunch allies Denise de Vries, Angela Ndalianis, and Helen Stuckey, who have accompanied me on the wild ride that is research; close friends and colleagues Karen Vered, Katharine Neil, and Jo-Anne Duggan; Gonzalo Miranda, who understands the wisdom—and pitfalls—of taking things apart to see how they work; and my biggest fan and most unwavering supporter, Dennis McDermott, who sadly missed seeing the revised manuscript being sent off by a matter of weeks. This book is dedicated to our daughter, in whom the experimental ethic is strong.

Melanie Swalwell

Melbourne, Victoria

1 Introduction

> Today, 2010, we can think of the computer gaming world as a huge funnel. Entering the top are hundreds of millions of game players. Some are way more serious than others and play practically every waking hour, but they are basically game players. As we get down to the spout, we find that a small number of players are becoming programmers. What does it take to go from being a player to being a programmer?
> —David Ahl

Today, gaming is a pastime enjoyed by large numbers of people, yet many games are now so complicated that only a few can actually program them. It hasn't always been that way. The order of the question in the preceding epigraph—from the foreword to the 2010 edition of David Ahl's famous book *Basic Computer Games*, which was one of the first books to gather game programs for users to type into a computer—regarding what it takes to go from being a player to a programmer has been reversed (Ahl 2010). When the collection was first published in the 1970s, it was the other way around: if one wished to play a game, it was necessary to first type in the program, whether the code came from a book or magazine or was of one's own devising. Many of those who played games on computers prior to the availability of removable storage media were programmers first and players second; indeed, some enjoyed the programming challenge so much that they did not bother playing the games they wrote.

Low-end, 8-bit machines offered many their first taste of computing, and many people wrote their own programs, experimenting with what it was possible to code on the then new micros, as they were called. Small and inexpensive compared to their minicomputer siblings, the arrival of 8-bit microcomputers—systems such as the TRS and System 80s and the Sinclair, Atari, Microbee, and Commodore ranges—heralded the moment when computers came within reach of laypeople. I refer to the years when these 8-bit microcomputer systems predominated in households—roughly the late 1970s to the mid-to-late 1980s—as the *micro era*. The development of games by users for 8-bit micros—termed *homebrew games*—was a significant use of these computers. It was a time when it was possible for one person to envisage, plan, and execute a computer game, doing the graphics, the sound, and the coding themselves. This book traces the contours of this activity, arguing that it represents a significant chapter in the history of vernacular computing.

The proximity of the 1970s and 1980s makes this relatively recent history, and yet so much in computing has changed in the last thirty years that these computers and the practices their appearance made possible can seem quite foreign. The period charted by this book was a moment in computing very different from the present one. Without painting it as overly strange and exotic, it began at a time when Microsoft was still developing the DOS operating system, and the Macintosh graphical user interface (GUI) hadn't yet appeared. Indeed, the term personal computer (PC) was just entering the lexicon, and PC was not yet the default term for a computer. While the now well-known home computer brands of IBM and Apple have been available for purchase since 1977 (Apple) and 1981 (IBM), their early systems were expensive compared to the "all in one" microcomputers, most of which simply plugged into a television set. The era predated "user friendliness" in computing, which meant that people had to invest more in learning how to use computers; users arguably also had more control. Digital games were still new and had not yet settled into the patterns of global marketing with which we are now familiar. It was a time when it was still common to play games in the public spaces of arcades.

It was against this backdrop that enthusiasts around the world purchased (or built or otherwise procured access to) early microcomputers. This book focuses on the experimentation that resulted from users' encounters with these little computers in the specific contexts of 1980s Australia and New Zealand. It draws on interviews and archival research to consider the phenomenon of homebrew game development as a particular mode of engagement with the then new technology of microcomputing.

Defining Homebrew

I define homebrew game development as having five main characteristics. First, homebrew productions were made in domestic space rather than in any kind of institutional space. Second, the creators of homebrew games were largely self-taught programmers. Some would subsequently go on to make software development their profession, but it did not start out that way. Third, most homebrew games were made by one person. Fourth, homebrew games often were not published, but if they were, it was usually on a small scale, with distribution often confined to the local level. A fifth characteristic is that homebrew authors' productions were marked by an experimental ethic. Coding at home was a highly experimental activity. It was a moment when users were seeing what it was possible to do with computers. As such, homebrew is more appropriately considered on its own terms rather than by comparison to the then industry state of the art, though there have also been moments when homebrew came into proximity with the commercial industry, and it is instructive to consider these moments.

I advance these five characteristics—domestic location, amateur programmers, sole creators, local distribution, and experimental ethic—in the interest of developing a working definition of homebrew rather than as hard and fast criteria in any proscriptive sense. That there were a range of exceptions, caveats, and borderline cases is unsurprising given how diverse my informants' paths to homebrew were and how much their circumstances varied. Most were working in relative isolation and did not identify what they were doing in terms of a particular label other than in descriptive terms—for instance, knowing that they "wanted to write games." I acknowledge that my application of a label to these practices runs the risk of making them seem more coherent and intentional than they perhaps were. I have opted to use the term *homebrew*—alongside other descriptive terms, such as *home coding*—to reference these practices for three reasons. First, although the term *homebrew* was not used widely in relation to computer coding in the 1970s and 1980s, it was used. Of course, homebrew was the name of the Bay Area computer club with which such famous names as Steve Wozniak and Steve Jobs are associated.[1] But beyond this obvious reference, the *Oxford English Dictionary* (*OED*) indicates that building of both computers and software were referenced as "homebrew" activities, with the published examples of the term's use dating from 1977 and 1989, respectively. A second reason is that using the homebrew conjunction brings out the rich resonances between the act of writing software and the preparation of other DIY brews in the home (e.g., beer, food, and other items), something I consider in detail in chapter 3. Finally, the term *homebrew* creates a link between historic game development practices and the current fascination with game production for a range of contemporary and historic platforms, including the 8-bit micros discussed here.

Communities are once again proliferating around 8-bit micros and consoles, sharing their creations for these vintage platforms via the full range of contemporary online media, and this is a topic that a history of homebrew needs to consider. This is not, then, just a book about the historic past, closed off in some timeless realm from the present. Consistent with the move to undertake new media histories and excavate media archaeologies, this is a book about both the past and the way it inflects the present. I am committed to imagining the present in "a dynamic perceptual relationship" with the past, as C. Nadia Seremetakis (1996, 4) so nicely puts it. While there is no equivalence between the 1980s and now, the appearance of texts such as Anna Anthropy's *Rise of the Videogame Zinesters* (2012), inexpensive computers such as the Raspberry Pi, coding tools such as Visual Basic, and the reintroduction of coding in school curricula are once again sending the message that anyone can write code and develop simple programs and games in ways that evoke the early micro era.[2] It would be remiss of a book on homebrew game development *not* to consider the significance of these developments in proper historic context. That said, historians are rightly cautious of the oversimplified use of history in so-called presentist accounts, which interpret past events anachronistically by appealing to contemporary attitudes and values. Such appeals to contemporary phenomena are often viewed as a search for relevance. This book is not doing that. Rather, I argue that by carefully considering historical practices and discourses of microcomputing in the 1980s, we can see how aspects of these inflect the present moment, and we can recognize resonances and differences between the past and the present. An appreciation of the history of homebrew game development should historicize—and help to complexify understandings of—current phenomena such as the interest in retro game development and moves to institutionalize the teaching of "computational thinking" in schools.

An Overlooked History

This book is dedicated to considering the history of homebrew game development in the 1980s. That the subject has so far not been broached is remarkable; I say this while also acknowledging the recent work of my colleagues in game history, including in this series and a trifecta of books published in 2016 on British game history that mention homebrew (Švelch 2018; Lean 2016; Gazzard 2016; Wade 2016). Given this, it may be more accurate to say that—at least until recently—homebrew was deemed an unworthy topic for study. Prior references have been largely dismissive, as if there is nothing interesting to be said about the topic. These attitudes have come from a number of different directions.

First, software historians and other commentators have frequently characterized hobbyist creations as amateurish. For instance, Martin Campbell-Kelly observes that: "The lack of significant barriers to entry led to the phenomenon of the 'bedroom coder.' Thousands of would-be software tycoons began to write games in their spare time, selling their programs through small ads in computer magazines. The typical game cost $15 and consisted of a smudgy, photocopied sheet of instructions and a tape cassette or a floppy disk in a plastic bag. Most of the programs were disappointing to their purchasers" (Campbell-Kelly 2003, 277).

It's hard to know what evidence this claim that "most of the programs were disappointing to their purchasers" is based on; perhaps Campbell-Kelly played a lot of games. Notwithstanding this possibility, it is worth noting that there is often a rush to make pronouncements on the quality of homebrew games.[3] It does not follow that simply because games were produced by amateurs, some of them weren't good games. Other scholars have treated the generation of homebrew game content as an early stage in the reception of computers, one that is presumed to have been quickly left behind as people moved on to playing (presumably more sophisticated) commercial games (Veraart 2011). Writing off homebrew games in this fashion misses the significance of the fact that laypeople—most of whom had never programmed before—were writing code and developing software. Nor is the homebrew phenomenon adequately accounted for by the view that it was just the training ground for the "real" business of professional game creation (Camper 2008, 151).[4] To treat homebrew game development in such a functionalist way again misses its wider significance as a mode of engagement with a then new technology. Tellingly, the dismissive accounts seldom pay any attention to the actual games themselves or to their reception. The implication is that—in agreement with Campbell-Kelly—the games were mostly uninteresting and disappointing, just copies of other games, and not worthy of attention. The field of cultural studies has alerted us to the cultural and political work that is done by such a discounting of popular everyday practices.

My aim in this project is not simply to recognize, revalue, and reclaim legitimacy for the formerly overlooked and disparaged activity of homebrew game development. I also seek to connect game history with the rich cultural and media theory around everyday life, as well as with critical perspectives on user-generated content. One of the reasons why homebrew matters and warrants remembering is because it provides a case study of the moment when everyday users first had the opportunity to create cultural products digitally. I argue that these nonexpert users brought quite different perspectives to the microcomputer and that studying them enriches our understanding of what was effectively the beginning of a vernacular digitality.

A second, related reason for dismissing homebrew game development stems from an impression that homebrew developers lacked professionalism as some (perhaps) moved into the world of entrepreneurial business. There is more going on here than initially meets the eye.[5] Some in the then nascent game development industry actively circulated this very self-serving impression. Marketing guy Gerry Gerlach, quoted in a 1987 profile piece in *GEM* magazine, perhaps inadvertently let on what was at stake in keeping the amateurs away: "Getting into that [publisher's] door, there's no question, there is a large amount of luck as well as knowing where to go. The barriers are just too great, there's just so many people wanting to do it. Particularly now that the English market has opened up into the American market, there's four square million kids writing games and sending them off on a daily basis. Having a track record or a decent referral makes a lot of difference" (Gerry Gerlach cited in Farrell 1987, 11).

While having an inside track might have helped these self-proclaimed industry insiders, it was not necessary to have a publisher, because there were a range of informal distribution channels open to homebrew developers, at least in the first half of the 1980s, and they took advantage of these "shadow economies" (Lobato 2012). Many dealt directly with their players, self-publishing, using mail order, or negotiating with third-party distributors. But the tendency to trivialize homebrew developers' professionalism is also more than slightly ironic, as many big-name developers effectively started as homebrew developers, including such well-known names as Richard Garriott (Origin Systems), Ken and Roberta Williams (Sierra Online), and Scott Adams (Adventure International). These developers all started selling their software in very low-tech ways, including in the proverbial Ziploc bags (King and Borland 2003, 38; Donovan 2010, 56–62). One wonders whether there is not a kind of embarrassment attached to this fact in some quarters: that somehow the beginnings of the industry that is now such a behemoth—bigger than Hollywood, as the claim routinely goes—should be more impressive, more professional than this. Stories of children beavering away at home after school—which is how many of the games I discuss came into being—are probably not seen as the stuff of origin myths. Nevertheless, homebrew developers operating under such ordinary circumstances deserve recognition, even if this is unlikely to come from industry.

A further reason why homebrew development seems to have been overlooked is because the low-end microcomputers that most homebrew developers used have themselves largely been ignored in the history of games and history of computing. The lack of scholarly accounts of the reception of microcomputers and their use by nonspecialists is a major gap in scholarship. To date, historians of computing have tended to focus on "big iron," unique computer systems with their own names (ENIAC, UNIVAC, and, in my part of the world, SILLIAC and CSIRAC). Thomas Haigh acknowledges that "the development of early one-off computers [is] one of the best documented corners of the history of information technology" (Haigh 2011, 448). When micros were new, they were frequently dismissed as toys by serious computer people, some of whom were also not keen to be associated with the (in their eyes) childish activity of gaming. One wonders whether such attitudes continue to be a factor in perpetuating a bias against these little, mass-produced computers. There is at least a degree of elitism around the popularity that micros enjoyed, as Haigh acknowledges in pointing to a generational change in scholarship: "At one time or another Atari, Texas Instruments, and Radio Shack held significant chunks of the American home computer market but none seems to have received significant historical attention. Commodore, which produced the bestselling computer model of all time (the Commodore 64), has been largely ignored by historians and journalists alike because its breakthrough hit was an inexpensive machine for home use. . . . The first substantial work on the history of personal computing is now arriving in dissertation form from a new generation of scholars" (Haigh 2011, 452).

While there is some extant research on micros (Haddon 1988; Veraart 2011; Kirkpatrick 2007; Turner 2006; Friedman 2005; Lean 2016; Sumner 2012; Saarikoski and Suominen 2009)—much of it related to games—that such a popular and widespread phenomenon as home computing should have received so little scholarly attention is surprising. Things are changing, slowly, and this book builds on recent scholarship on microcomputers and microcomputing by authors such as Tom Lean (2016), Alison Gazzard (2016), Jaroslav Švelch (2013a), Graeme Kirkpatrick (2015), and others. Fans have also taken up the slack, organizing and publishing the histories of their beloved systems online (e.g., Lemon64, World of Spectrum) and in book form (Hague 2002; Dyer 2014; Wiltshire 2015).

Not only has computer history managed to ignore the best-selling computer of all time, Haigh acknowledges that it has remained uninterested in the directions taken in other branches of the humanities, such as the so-called cultural turn of the 1980s. He writes that "most work in the area continues to reflect the mindset and concerns of technical specialists rather than the concerns of academic historians. Its questions and methods are at best unconcerned with academic fashion and at worst antediluvian. . . . Histories of information technology have rarely considered representative experiences, social changes, or the influence of information technologies on different kinds of work. To be blunt, outsiders from more mature historical subfields are likely to find the bulk of existing scholarship narrow, dry, obsessed with details, underconceptualized, and disconnected from broader intellectual currents" (Haigh 2011, 471).

The challenge of connecting computing history with the popular, with user practices, and with the cultural study of technology's reception is one I take up in this book, and in a moment I will detail how I do this. But first, I need to note that, just as micros and the homebrew phenomenon have been trivialized and overlooked by aspects of the game industry and computer historians, mainstream game history has also largely overlooked microcomputers, particularly (and somewhat surprisingly) in the US. To date, game history has been noticeably arcade and console heavy (Kent 2001; Burnham 2003; Loguidice and Barton 2009; Montfort and Bogost 2009; Guins 2014; Kocurek 2015) and light on mentions of microcomputers, with only a handful of exceptions.[6] In part, this is probably because game history itself has focused on the popular, and mass-market consoles trump micros on sales figures alone. As King and Borland note in their book, "At virtually all times covered . . . sales of video games for home console platforms such as those made by Atari, Nintendo, Sega, Sony, and Microsoft far outstripped most of the computer games we are writing about" (King and Borland 2003, 6). It's also been argued that the microcomputer scene was "geekier" (7), with personal computers being culturally coded as more "serious" machines (Kline,

Dyer-Witheford, and De Peuter 2003, 142) and therefore more specialized than the mainstream, mass-market audience for consoles (Kirkpatrick 2015). The homebrew scene is more accepted within the UK's game history, where it was a sizable practice and has been acknowledged as an important incubator of design and programming talent (Lean 2016; Wade 2016; Gazzard 2016). The documentary *From Bedrooms to Billions*, for instance, includes mention of phenomena such as people selling games out of the backs of their cars. While the relative dearth of scholarship on games for micros in the US is puzzling, it is clear that enthusiasts there also developed homebrew games and other software for microcomputers. In addition to the accounts of fans such as Kevin Savetz and Rob O'Hara, there is ample archival evidence. Not only have I found homebrew titles while trawling through the Stephen M. Cabrinety collection at Stanford University, but the early West Coast Computer Faires were billed as "exclusively devoted to home and hobby computing," servicing what was clearly a lively hobby computer scene (Warren 1977; Warren 1978; Alpers 2014). While Britain had *The Computer Programme*, a television show produced by the BBC that began in early 1982 and auspiced the BBC microcomputer (Gazzard 2016, 39–42; Lean 2016, 102–106), the US had *The Computer Chronicles*, a PBS television show that aired continuously from 1983 to 2002 (nationally from February 1984). I have seen William Shatner promoting the Vic-20 in television advertisements posted to Facebook, just as John Cleese did for the Compaq in the UK and John Laws for the Commodore 64 in Australia. Finally, there were plenty of dedicated US computer and game magazines (such as *Dr Dobbs Journal, Kilobaud, Softalk, Compute!,* and *Computer Gaming World*) and user group publications that featured type-in code for games.[7]

Theoretical Framework

In this book, I connect homebrew game history with perspectives from cultural and media studies, particularly the rich vein of scholarship on everyday life, consumption, and reception. The contribution of cultural and media studies to the study of users and consumption is well recognized, even outside these related disciplines. In their introduction to *How Users Matter*, for instance, Nelly Oudshoorn and Trevor Pinch write that "scholars in the fields of cultural and media studies [have] acknowledged the importance of studying users from the very beginning. Whereas historians and sociologists of technology have chosen technology as their major topic of analysis, those who do cultural and media studies have focused primarily on users and consumers" (Oudshoorn and Pinch 2003, 11–12).

Although the term *user* has sometimes been critiqued for its drug connotations (Neumark 1993), it is now in common usage, understood to refer to "a person who uses or operates something" (*OED*), and within media studies as specifically one who uses interactive technology. The territory of user studies encompasses scholarship drawing on the popular, the ordinary, the vernacular and quotidian, and the folkloric. Whereas cultural and media studies might be better than other disciplines at studying users and consumption, scholarship on what users did with computers in the micro era is still not well developed. In game studies, contemporaneous ethnographic research has been conducted with players since the mid-1990s, demonstrating the richness of this method for understanding what players do with and around games; however, not only has the emergence of historic game studies been more recent, but the moment for conducting contemporary ethnography with micro users has passed. In this book, I seek to demonstrate the importance of theoretical traditions such as fan and audience studies and media and computer histories to the emerging subfield of game histories and vice versa.The considerable slippage between terms such as *audience, user, consumer,* and *fan* (and the relatively low frequency with which terms such as *enthusiast* and *hobbyist* appear)—while explainable in terms of debts to earlier communication models and theories—presents an invitation to reconsider both terminology and the historicity of the user. Editors Olia Lialina and Dragan Espenschied, for instance, dedicated the title of their book *Digital Folklore* to computer users in a specific period of vernacular digitality, that of the early web. Lialina and Espenschied assert that by the 1990s home computer culture had ceased to exist. They claim that by then the web was populated by "Real Users" and "Naïve Users," with user becoming a derogatory term for those who showed up on AOL with no technical skills, needing things to be as simple as possible (Lialina and Espenschied 2009). This book might be thought of as a prequel to that era, but it is fair to say that the predominance of computer industry history has led to a neglect of what users have done. Sometimes, industry histories include a focus on what users are supposed to have done or what it is assumed that users do, but this is not the same as what users actually do and did. Discourse sets norms, but user practices need not adhere to such norms; practice exists in a social context shaped by discourse, yet it can diverge from that discourse. Users *can* be unruly and do their own thing, though they may not be. Discursive analyses on their own are not enough to recover what users did with technological products.

The theorist whose work I find most helpful in thinking about early computer users and the products of their use is Michel de Certeau. In *The Practice of Everyday Life*, de Certeau memorably writes that we know little of the uses that people make of things; that is, what consumers actually *do* with products. He makes a distinction between production and the uses that are made of products by "users who are not producers," treating the latter as a sort of generative consumption. Conceiving of consumption as a form of production is helpful, if initially paradoxical, as it facilitates inquiry into practices often passed over because they are not immediately obvious. As de Certeau writes,

> The "making" in question is a production, a *poiēsis*—but a hidden one, because it is scattered over areas defined and occupied by systems of "production" . . . and because the steadily increasing expansion of these systems no longer leaves "consumers" any *place* in which they can indicate what they *make* or *do* with the products of these systems. To a rationalized, expansionist and at the same time centralized, clamorous, and spectacular production corresponds *another* production, called "consumption." The latter is devious, it is dispersed, but it insinuates itself everywhere, silently and almost invisibly, because it does not manifest itself through its own products, but rather through its *ways of using* the products imposed by a dominant economic order. (de Certeau 1984, xii–xiii, italics in original)

The verb *faire* in de Certeau's French subtitle *Arts de Faire* communicates the sense of active creation (in French, *faire* means to make or to do) more effectively than the English term *use*, which tends to imply functionality and instrumentality.[8] In volume 1 of *The Practice of Everyday Life*, de Certeau details these insights on consumption by referring to such diverse examples as what the cultural consumer does with televisual images, the use of urban space, the products purchased in the supermarket, stories and legends distributed by the newspapers, wandering, renting, reading, and speaking. But in this first volume he is largely preoccupied with writing and reading, and many of his points do not transpose well to a more general concern with consumption in everyday life. It is in the second volume—a book written by de Certeau and his collaborators Luce Giard and Pierre Mayol—that the everyday literal practices of consumption, production, and making are considered, via case studies of shopping, cooking, and the inhabiting of neighborhoods (de Certeau, Giard, and Mayol 1998). The microethnographies present in-depth accounts and analyses of the everyday practices of "ordinary" French villagers. It is to these studies that I will turn in chapter 3 to help articulate the significance of homebrew game development as a key moment in what I term a *vernacular digitality*.

Vernacular Digitality

As already stated, the era of 8-bit microcomputing marks the moment when computers first came within reach of laypeople. It was a reasonably egalitarian moment: while access certainly was not afforded equally or to all, computers were, for the first time, available to anyone interested who had $300 or so to spare. The adoption and use of these first-generation home computers marks the beginning of a vernacular digitality. Following de Certeau, this study is concerned with early computing and computer culture as it was practiced by "ordinary" people, not what was valued by "officials" (de Certeau, Giard, and Mayol 1998, 251).

Games had a central role in the reception of computers in the home, more so, I will argue, than many other types of software. Games were one of the biggest uses for these early computers, if not *the* biggest use. Players learned—perhaps first in the arcades—that onscreen movement happened in response to their input, but games on micros acted to make the new digital technology familiar. Games literally domesticated computers for users, yet the history of homebrew game development I recount in this book has been a hidden history. This is partly because the reception of microcomputers occurred largely out of public sight. The private space of the home has traditionally been devalued in comparison with public space (Lloyd 1984), so besides marginalizing amateur perspectives, there are issues of legitimacy around the domestic, the everyday, and the hobby in play here. As Elaine Lally's book *At Home with Computers*, on users' engagements with computers in the subsequent decade, has shown, these practices are not typically found in the archival record. Neither public nor spectacular, they require participatory methods (she used ethnography) to uncover (Lally 2002).

Two further factors have kept homebrew in the shadows: its everydayness and a lack of certainty about how we should talk about it. Many of my informants acknowledge that they didn't talk much about what they did. Sometimes this was because they didn't think it was very remarkable—they felt that everyone was doing it—while others feared that their coding activity wouldn't be understood by their peers and friends and that they would be stigmatized because of it. Finally, we still lack a discourse beyond nostalgia for the recuperation of the early microcomputer period in the contemporary moment. In part, this is a larger problem with histories of computer and game culture, as I've argued elsewhere (Swalwell 2007). It is as if we haven't quite known how to contextualize and critically situate this period in social and cultural history, though its significance as the moment when digitality entered everyday life makes the importance of doing so undeniable. This means that, until now, the distinct perspectives of homebrew game creators had largely been unavailable to the history of games and of computing more generally. Developing such a discourse—in which the novelty and significance of homebrew practices may be appreciated within and beyond this field—is therefore another goal of this book.

Game History

In the absence of many contributions from scholarly game historians, a game history has grown up that has been dominated by insiders and journalistic accounts, what Erkki Huhtamo has termed game history's "chronicle era" (Huhtamo 2005, 4). Histories published around the turn of this century were typically told with a focus on the US or Japan and tended to assume the uniformity of products and reception worldwide. Central debates were concerned with what "great men" (e.g., Ralph Baer and Nolan Bushnell) did, while a range of foundational stories and markers (the shortage of 100 yen coins following the Japanese release of *Space Invaders*, the dumping of *E.T.* cartridges in the desert, "the" video game crash of 1983) established historical narratives and major turning points in the industry (Kent 2001; Herz 1997; DeMaria and Wilson 2002; Burnham 2003; Gielens 2000). A little later, Tristan Donovan's *Replay: The History of Video Games* (2010) attempted to tell a history of software rather than just hardware. While his book is still a journalistic account, Donovan recognized that the history of games is global rather than solely North American. One chapter covers game development in 1980s Britain, Spain, and Australia, and another considers games on microcomputers. Amid the welcome recognition that regional variations existed, however, was the treatment of audiences as passive, a point that rankles critical media scholars.

From the early search for origins, stories of great men, and attempts to write the definitive history of games, scholarship with a greater concern for specificity, plurality, and inclusivity has been emerging as the game history field undergoes processes of refinement and maturation. With this comes a greater commitment to social and cultural histories of games and gaming, a tighter focus on particular regions, such as the former Czechoslovakia (Švelch 2018); issues, such as LGBTQ (Shaw 2019); genres, such as wargaming (Harrigan and Kirschenbaum 2016); or players, as well as a cognizance of, and engagement with, theoretically informed historical inquiry (Lowood and Guins 2016; Swalwell, Stuckey, and Ndalianis 2017).

To such departures from standard digital game history narratives, this book adds one more in that it explicitly foregrounds the "where" question. Where did game history take place? From which locales are game histories told? Digital game history did not unfold evenly or uniformly across the globe, and the particularities of space and place matter. Yet most digital game and software histories are silent with respect to geography. The orthodoxy that the US, Japan, and—to a lesser extent—the UK constituted the centers at the outset of the industry has enjoyed such legitimacy that many accounts do not even bother to mention where their material or statistics pertain to (e.g., Campbell-Kelly 2003; Montfort and Bogost 2009). That many histories have largely accepted the game industry's global rhetoric has no doubt contributed to this situation. This means that locality and difference were largely left out of game history, at least until recently, when more diverse case studies began to be published (e.g., Fassone 2017; Švelch 2018). Sometimes, exhibitions have preceded published scholarship, as happened with the Berlin Computerspiele Museum's early interest in game history around the world and more recently the Finnish Museum of Games' in-depth treatment of games and game reception in that nation (Nylund 2018).[9] Given the great historic diversity of games and contexts for their play, it is important to develop an appreciation of sociocultural and geographic specificity.

This book presents a critical history of homebrew game production in the specific contexts of 1980s Australia and New Zealand, a region far from the perceived centers of game development but one with its own remarkably rich development histories (see, for instance, Swalwell and Davidson 2016; Stuckey 2013; Stuckey and Swalwell 2014; Swalwell and Loyer 2006; Swalwell 2015; Brown 2003). In this book, microhistorical[10] case studies bring a specificity and a richness of illustration to the argument, as do the semistructured interviews with informants. Many of the archival sources and examples I use—from magazines to microcomputer brands—are unique to the region. Fostering inclusiveness in (game) histories is a worthy goal, and Australasia is self-evidently on the periphery. With this study, I hope to reintroduce some of the actors, sites, technologies, products, and practices that have been left out of existing accounts of game history. However, while a local emphasis apparently encourages specificity, it can masquerade as yet another version of the search for origins (Foucault 1984). I recognize the need for this kind of microhistory to be able to scale to ask larger, more general questions, as Giovanni Levi (n.d.) puts it. I want to do more than simply appeal for the inclusion of minority histories and discourses in majoritarian accounts. I seek to offer some answers to questions such as how we might write local game histories and how one might position the local without falling into the trap of exceptionalism. These are pressing questions for those researching histories not of the centers of game development but rather its peripheries.[11] A scholar with a local focus will likely be asked to set their work in a wider context to make its significance clear to nonlocal audiences, a burden not demanded of those writing from the center. This raises the prospect that doing local game history might actually mean comparative game history, a point that demands serious consideration.

Heterodoxy

While it is a common complaint that New Zealand often gets left off the map (figuratively and literally), my use of this local material is not a simplistic attempt to replace a global or US-centric history with a local one. For one thing, when undertaking game history research in Australasia, it quickly becomes evident just how imbricated the local and the global are, something I've discussed at length elsewhere (Swalwell and Davidson 2016). In the 1980s, people in this region were in touch with, yet also distant from, all the major centers of game development and consumption. The local then is (almost) always heterogeneous, already imbricated with global or nonlocal elements, something seen in some other recent game histories (e.g., Fassone 2017; Švelch 2013). My approach to writing local game history is *heterodox*: it not only undermines many of the orthodoxies of game historiography I outlined earlier but, as I will argue, research from the "periphery" can also disturb what we thought we knew about the "center." I borrow the term *heterodoxy* both from Italian microhistorian Carlo Ginzburg and from de Certeau and Giard's short essay "Ghosts in the City," in which they write about the corners of Paris that escaped urban renewal (de Certeau and Giard 1998b; Ginzburg 1992).[12] The "old things" de Certeau and Giard discuss manifest heterodox qualities in that they are not in keeping with the modern city–they hold the "ghosts" of the past, not just memories but also what has been forgotten–and the tastes of the people who inhabit these places are often at odds with those of urban planners. Though they are discussing the built environment, there is also a strand in their thinking about "old fashioned thing[s becoming] incorporate[ed] in the national heritage," and it is this theme of the heritagization process and the politics of national heritage that resonate for me with a game history concerned with the local, as it highlights the work that is done in the name of saving national heritage.

Invoking the salvation of heritage under the sign of the local can be a powerful argument, and sometimes it is necessary to be polemical in the face of indifference; digital games are, after all, one of the most illegitimate media forms, made on one of the least esteemed cultural platforms (the computer, with micros often regarded as toys). But while invoking the salvation of local heritage is an understandable tactic, I argue that there's a need to avoid the pitfall of local (or national or regional) exceptionalism, because what is involved is little more than a substitution of one set of heroes for another. To research the first game company from country X or the first successful game from country Y can be just another version of the search for origins. If in game historiography the formulaic recitations of firsts and great men and the view from the center represent unhelpful and undesirable orthodoxies—as I've suggested—then merely substituting or supplementing them with the view from the periphery achieves little. Nor does valorizing the peripheral for its own sake—as amusing as its oddities and quirks may be—achieve much. Game history can be more than celebratory histories written by insiders or obscure and arcane facts assembled on specific platforms. We need to find our way out of these old orthodoxies without simply replacing them. In this book, I ask what such a heterodox historiography might look like.

As this book will demonstrate, game history can address ordinary people's experiences as they came to terms with the new digital technology of computers. The dislocation that comes from addressing forgotten subject matter at odds with orthodox approaches to game history makes it possible to ask new and different questions of both the historical past and the present. Focusing on the relatively unknown phenomenon of homebrew development serves to move the coming of the personal computer narrative out of the realm of the familiar and what we think we already know about it. The view of homebrew from the periphery effectively defamiliarizes what has been until now a largely US-centric game history so that questions as to why there has been so little attention paid to histories of micro use in that nation might be asked more often by scholars.[13] Ironically, decentering what Corinna Schlombs (2006) has identified in computer history as the North American default perspective might actually help to stimulate inquiry into microcomputer reception in this large and important market.[14] In this sense, the project can be seen as sharing some of the aims of media archaeology, as Huhtamo has articulated this: "Media archaeology means for me a critical practice that excavates media-cultural evidence for clues about neglected, misrepresented, and/or suppressed aspects of both media's past(s) and their present and tries to bring these into a conversation with each other" (Huhtamo 2011, 28).

Other questions that the study enables to be raised include ones particularly around audience and disciplinarity. I have already presaged that 1980s homebrew provides a historical context for the contemporary retro homebrew and indie scenes. I also argue that the period and practices of homebrew enable certain questions about audience in twenty-first-century digital culture—particularly concerning the productivity and making of users—to be reapproached. Many writers and theorists have attended to audience or user production, devising various labels for the activities of fans, hobbyists, readers, enthusiasts, and others, including "prosumers," "Pro-Am" divides, "gleaning," and the like (Marshall 2004; Jenkins 1992; Bruns 2008; de Certeau 1984; Gruber Garvey 2003; Fuller 2012). The phenomena of appropriation, modification, and remixing have long been central topics within media and fan studies, while modding has been a subject of considerable interest within game studies since at least the late 1990s (Sotamaa 2009; Morris 1999; Schleiner 2002; Champion 2012a).[15] Indeed, it is fair to say that thinking of audiences as productive has become commonplace, culminating with the nomination of "You" as *Time* magazine's person of the year for 2006 for the role of users as generators of Web 2.0 content (van Dijck 2009). While the advent of Web 2.0 raised the profile of digital media consumers' productivity in a digital age, too often these aspects of an audience's cultural practice are presented as if they are without precedent. Despite a history extending more than thirty to forty years, one could be forgiven for thinking that such consumer practices only began with the introduction of the internet, so little attention has been paid to pre-internet practices of digital consumption. Charting 1980s homebrew provides a sort of prehistory of participatory digital media practice, offering the chance to reframe contemporary audience and user practices in longer arcs of cultural history. My excavation of the early computer user is intended to contribute to a wider media archaeology of user production (Huhtamo and Parikka 2011), where this is understood as "uncover[ing] and circulat[ing] repressed or neglected . . . approaches" and "lost traces . . . normally . . . occluded by more obvious narratives" (Parikka 2010).

I believe the time is ripe to revisit early computer culture. On the one hand, the computer industries continue to espouse linear narratives of progress. On the other, the near ubiquity of computing in today's almost always on, almost always connected world means that we are rapidly losing touch with the moment of transition between analog and digital ways of living. While computers continue to be identified with new media, revisiting the time—more than thirty years ago—when computing was new and exploring the engagements of users with early microcomputers destabilizes an overly simple identification of new media with the future. One line of inquiry this scholarship facilitates concerns the "informal" hardware hacking practices that 1980s computer enthusiasts routinely engaged in. The rise of laws to ostensibly protect intellectual property—for instance, prohibiting circumventing of technological protection mechanisms (TPMs) and restricting the sale and use of mod-chips—effectively means that some common 1980s practices are now outlawed (Gillespie 2009; Schulz and Wagner 2008). Contemporary legal restraints effectively "black box" technology, so it is an opportune time to revisit a moment when users were free to read, write, repair, and modify the hardware they owned, freedoms that the contemporary open hardware and right to repair movements are working to get onto cultural and political agendas.

Interdisciplinarity, Methods, and Sources

This is a thoroughly interdisciplinary book, situated at the intersections of cultural and media studies, philosophy of technology, history of computing, and game history. The study is also informed by, and speaks to, the broader and more diffuse subfield that is concerned with conceiving and theorizing media audiences and creative practices, including fan studies. The main theoretical framework derives from de Certeau, but I also have cause to draw on philosophers of media and technology, including Walter Benjamin (on copies) and Raymond Williams (on experimentation).

My approach employs a mixture of methods, primarily archival sources and semistructured interviews. Ironically, despite my subject matter being digital, most of the archival sources are not, which has entailed the need to immerse myself in published and ephemeral archival collections. Often, judgments made in the past about what would be important in the future have not favored the topics I'm studying. The lack of publicly held resources and the recentness of this history have necessitated that I generate primary sources through oral history interviews as well as drawing on personal archives held by homebrew creators themselves. Some of the research has been done in the context of a game history and preservation project that I lead called Play It Again. Funded by the Australian Research Council, the project ran from 2012 to 2015 and focused on documenting and preserving artifacts—both analog and born digital—and memories of the early microcomputer period, centered particularly on games that were written for microcomputers in 1980s Australia and New Zealand. We collected games and artifacts, interviewed developers, and invited players and other members of the public to reflect on the period from a range of different perspectives, uploading their memories to a "Popular Memory Archive" (Stuckey et al. 2015; de Vries et al. 2013).

I will begin to introduce my informants in chapter 2, but first I want to comment on the interview component of the research. For some time, my preferred approach has been to record reasonably in-depth interviews with informants. Given the circumscribed nature of the topics, these are roughly an hour in duration and often more like conversations in which I prompt informants to unpack an area in detail, sharing their expert knowledge with me. As homebrew is an overlooked area not usually accorded attention, hearing the voices of homebrew developers is itself significant, and taking an interest in their knowledge and what it means to them yields extremely rich interview material. Interviewees' associations are often surprising, so I allow considerable freedom to follow tangents and digressions. Interviews are frequently in informants' homes and complemented with other forms of interaction— for instance, being shown artifacts—giving the method a quasiethnographic component.[16] Like Luce Giard, I am a believer in, and beneficiary of, the great "richness of speech among ordinary people if one takes the trouble to listen to them and encourage them to express themselves," and I use and foreground my informants' explanations extensively throughout the book (Giard 1998a, xxviii–xxix; de Certeau, Giard, and Mayol 1998, 160). Many informants have shared archival sources with me, and the book includes scans and analyses of a number of artifacts from the time. My interviewees are not just expert informants but also collaborators of a sort, and their perspectives enrich the text immensely.

Chapter 2 sets the context by considering the public discourses surrounding microcomputers and programming by home coders in Australasia from the late 1970s to the mid-1980s, drawing on published sources. It focuses on what appear to have been persistent doubts about the usefulness of microcomputers among the general population at the dawn of the micro era, questions that hobbyists and enthusiasts ultimately ended up getting drawn into and weighing in on. While Alex Wade has expressed surprise at how little literature of the time explored what he calls the "bedroom cultures" of coding (Wade 2016, 59), we do have the opportunity to examine public utterances and discourse on the uses of computers as these were aired at the time in books, magazines, and ephemera. There was considerable discussion of users inventing uses for micros according to their avocations, and while studying traces of published discourse is not the same as studying practice, I intersperse examples of practice with some of the published discourses to begin to highlight their divergence before diving into user practices in greater detail in chapter 3.

Chapter 3 takes up user practices and consumption, influenced by French theorist Michel de Certeau. Drawing on semistructured, in-depth interviews with users, I hone in on some of the features of homebrew game development. My thesis is that the early microcomputer users who wrote their own programs at home—those people I've elsewhere dubbed home coders (Swalwell 2008a)—are a strong example of de Certeau's insight that users and consumers are makers and producers of culture. I borrow insights from de Certeau's collaborators Pierre Mayol and Luce Giard on aspects of ordinary culture in their studies of provincial French life and cooking to draw out some of the characteristics of homebrew practice as a form of ordinary culture. De Certeau and Giard's articulation of the political, aesthetic, and ethical dimensions of ordinary culture are useful in assessing the significance of homebrew practice, offering the chance to highlight some of the rewards practitioners cite, such as the sense of satisfaction they derived from their activity.

Chapter 4 delves into how homebrew developers understood their practice and how they articulate their motivations. It addresses where authors' ideas came from and the influence of forms of popular culture, and argues that it is important to understand the gaming "ecosystem" in which many users were located (i.e., across arcades, consoles, handhelds, and micros). I offer a detailed case study of Nickolas Marentes's game *Donut Dilemma* as an example of how ideas for games formed and were developed. I then engage with a number of allegations that are made about homebrew games, including the charges that homebrew games were just "clones" and that writing games oneself was "just a stage." Finally, I ponder some of the decline theses that attach to microcomputing, specifically the claims Graeme Kirkpatrick and Frank Veraart make about computer programming as a hobby practice, situating these against the backdrop of changes in the computer market from the mid-1980s on.

Research into 1980s hobbyist practices exposes a number of features of computing in this era in addition to programming that are often passed over. In chapter 5, I present evidence that some users also built their own computers and were encouraged to hack their machines together with other hardware, developing various other new uses and functionalities. Users were involved with hardware in a way that few are today, but a culture of support—inherited from ham radio and electronics—helped such experimentalism thrive. Interestingly, the way that de Certeau's theories have been imported and used in cultural studies has largely focused on the work of fan labor and particularly the impact that fan communities can have on the official creative works that are released by film and television studios. This is perhaps not that surprising considering the heavy emphasis on writing and a scriptural economy in volume 1 of *The Practice of Everyday Life*. Homebrew gaming offers unique perspectives on these debates, and my use departs in several important ways from the now mainstream and quasicanonical understanding of users as productive. By this I mean the understanding that has developed in the dominant Anglophone reception of de Certeau, initially through John Fiske's claim that though "people cannot produce and circulate their own commodities . . . what they have are the products of the cultural (and other) industries. The creativity of popular culture lies not in the production of commodities so much as in the productive use of industrial commodities. The art of the people is the art of 'making do'" (Fiske 1989, 27–28).

Fiske's emphasis on de Certeau's "art of making do" and the way that "active audiences were able to oppose the dominant messages sewn into the media texts promulgated by media corporations" (Longhurst 2007, 8) was joined a few years later by Henry Jenkins's deployment of the metaphor of poaching in his book *Textual Poachers* (1992). Fiske's and Jenkins's perspectives were influential and spread widely through the then nascent disciplines of cultural, television, and fan studies. Homebrew presents an interesting limit case for cultural and fan studies because users did not just develop their own interpretations of industrial commodities or content based on popular texts; by using their micros, they literally made their own games. And beyond creating games, many users also built computers and tinkered with hardware. As I detail in chapter 5, electronics tinkering and hardware hacking were central to the early micro period, and there are disciplinary implications that arise from excavating these practices. Despite the attention that consumer productivity receives within cultural and media studies, one branch of the family tree of user production—that of tinkering with code, hardware, electronics, and engineering—has been more or less forgotten. Restoring the largely overlooked fields of electronics and engineering to the lineage of user making allows new and interesting questions to be raised about how user engagements with computers changed over the decades.

In chapter 6, I ask what the legacy of 1980s homebrew practice is and consider how this important period of transition to digitality will be remembered. The influence of 8-bit aesthetics and gameplay on the contemporary game industry, along with the ongoing practice of programming 8-bit microcomputers, is presented as evidence that micros are not "dead media." Rather, microcomputer practice persists. I argue that my informants' contemporary development of "demakes," together with software heritage projects, and the use of vintage games for contemporary political expression offer a new discourse on game history by bringing the present into a dynamic relationship with the past.

Finally, in chapter 7, I summarize the significance of the homebrew case study as the moment when computing first came within reach of ordinary people. I highlight some of the points of significance in the homebrew case study for media, cultural, and audience studies and computer history and identify some of the promising new research directions that follow from the study. Finally, I discuss some of the implications the study has for vernacular digitality and conceptions of digital cultural heritage.

2 Discourses about Microcomputers

The moment when microcomputers became available for home use was highly significant. Now, roughly forty years later, it is clear that personal computing has affected almost all aspects of our daily lives, including how we socialize and create culture. But in the late 1970s and early 1980s, few people outside research labs and those working in "the computer game" (as the industry was often called) had been given the chance to get their hands on a computer, let alone spend time with one in their home. Microcomputers entering the home marked the beginning of a vernacular digitality, if not yet the mainstreaming of computers. However, their embrace was not straightforward: doubts, suspicions, and a range of misconceptions had been inherited, often from popular cultural representations. This chapter addresses the moment when this now taken for granted technology was new by considering some of the discourses that surrounded the arrival of micros as recorded in published sources. Such discourses helped to shape the context in which homebrew game—and indeed other software—development took place. Consideration of discursive shaping prepares a foundation for considering the actual uses of computers by homebrew developers, which will be discussed in chapter 3.

I begin by reviewing perceptions of the usefulness of microcomputers held by users and nonusers, linking this to the uptake of computers within the home. The usefulness of microcomputers in the workplace is also considered briefly. I then discuss one of the biggest uses for early computers—the playing of games—and how games relate to perceptions of usefulness. I argue that computers suffered from a conceptual problem, that they were actually a technology in search of a use, and that programming was how uses could be developed. The discussion then turns to programming, considering some of the discourses found in the "teach yourself to code" books that were a central feature of the early micro era. A discourse of experimentation was significant here, and also in magazines, so I develop a framework to account for how computer users began to invent uses for computers. In a context where—apart from digital games—the range of software was limited and rather unimaginative, some hobbyists also began developing other forms of homebrew software. Such experimental use and creation presents an inflection point at the chapter's end, pivoting from discourses of use to anticipate the de Certeauian framework of "users making," which is expounded in detail in chapter 3.

Useful?

Today, the idea that the computer is a useful piece of technology is deemed to be so obvious as to not even be worth debating, but this was not always the case. As Graeme Philipson writes, "Many people who bought expensive and underpowered PCs wondered what to do with them" (Philipson 2003, n.p.). In this section, I critically examine some of the questions, responses, and ambivalence that attended the concept of early microcomputers' "usefulness." Discourses on computers' usefulness were recurrent in Australia during this period. Adopting utility as a lens allows me to examine several—at times interrelated—issues around the wider cultural reception of computers during this decade, including perceptions about microcomputers, doubts about their claimed usefulness, and users' invention of new uses for computers.

In the late 1970s and first half of the 1980s, low-end microcomputers were almost exclusively the domain of hobbyists or enthusiasts, those people who liked tinkering around with electronic gadgets or who enjoyed logic problems. Indeed, as Christina Lindsay notes regarding the Tandy Radio Shack computer, the inventors of this machine envisioned users who were "[electronics] hobbyists like themselves" (Lindsay 2003, 33). The prospect of owning a computer was exciting for people who were already interested in them. However, the wider population didn't really know what to think, at least in the Australian literature I've surveyed, and it took some time for that to change. While computers aroused fears and concerns in some about what computerization would mean, more often than not, the reaction seems to have been one of uncertainty, and from the published sources I've consulted, the main reason for this was not knowing what computers would be useful for.[1]

There are no statistics on computer use during this early period, as the earliest Australian Bureau of Statistics (hereafter ABS) data collection covering computer ownership and use is the 1994 *Household Use of Information Technology*. However, Ironmonger, Lloyd-Smith, and Soupourmas provide data on computer ownership from the middle of the decade.[2] In 1985, only around 6.6% of Australian households owned a computer (or 13.8% of households with children). In 1990, ownership was predicted at 15.1% of households (or 26.3% of households with children) (Ironmonger, Lloyd-Smith, and Soupourmas 2000). From these figures, we can say that in the late 1970s to early 1980s, hobbyists owning computers comprised only a small fraction of the population in Australia.[3] John Schmitt and Jonathan Wadsworth's summary of data from the British General Household Survey (GHS) and the US Consumer Expenditure Survey (CES) for the 1984–1998 period suggests that computer ownership was slightly higher in the UK (at 12.6% in 1985) than in Australia. The US figure was lower than for the UK: in 1988, the earliest year for which comparative data are given, 10.2% of US households owned a computer, compared with 17.2% in the UK. However, given that the wording of the question in both countries explicitly excludes games—a "computer, not solely for games" (CES) and "home computer (exclude video games)" (GHS)—the data need to be interpreted with caution in the present context (Schmitt and Wadsworth 2002).

What Were Computers (Supposed to Be) Good For?

As already noted, most hobbyists did not need to be convinced of the wonders of computing, but those around them often did. In this section, I ask what these microcomputers were said to be useful for. This is aligned with what they might have been used for (the focus of chapter 3), but it is not quite the same. While there was a reasonably widespread, commonsense expectation that computers should be useful, the question of whether early low-end microcomputers were actually useful generated quite a bit of discussion. Clearly, much doubt existed in some peoples' minds.

Why did people doubt the usefulness of computers? The expense of the purchase seems to have been a major reason. Though relatively cheap compared to a minicomputer, a micro could still require a considerable outlay of cash in the early days, depending on what was bought. In the 1980s, Katharine Neil—now a professional game developer—was an adolescent who dabbled with coding. The following excerpt from our 2006 interview helps to put the financial outlay of a computer purchase into context. Neil recalls, "I remember my parents saying, 'Well, we can either get a microwave, a video recorder, or a computer.' Those were the luxury items, the new luxury items, and if you wanted to be an up-and-coming household, you'd get those things, or one of them. So we chose a computer. But it was about $3,000 or something, outrageously expensive."

While Neil is speaking of the New Zealand context, owning a computer was not a necessity in either Australia or New Zealand in the 1980s. It was a discretionary purchase. Given the expense, it's not surprising that people wanted to know what it would be useful for, what it would do for them. Letters pages in magazines attest to the doubts some expressed. Spouses were irate that discretionary income was being expended on something they often didn't understand, and if one spent too much time at the computer, Wideman wrote, "Friends and relatives start to forget what you look like. Spouses and lovers take up other interests" (Wideman 1982, 90). For this reason, he counseled against the idea of buying a basic computer with the intention of adding on to it later: "A computer that doesn't really do very much, and consumes a large amount of your time, is even less well accepted by those associates or other members of the family who might have some say in further expenditure. . . . Don't just go for the cheapest starter outfit. Pay attention to the overall cost of a system expanded to the point at which it is going to be most useful, which inevitably includes software" (92).

I suspect the claimed uses for computers did not help much in swaying the doubters. In the early 1980s, all sorts of claims were made for what microcomputers would be good for in the home. These included such unlikely tasks as recipe filing, the preparation of household budgets, and scheduling auto maintenance. In 1978, Rudi Hoess—who would later be credited with introducing the Apple II to Australia (Hearn 2006)—claimed that microcomputers were changing the image of computers to "that of a friendly servant capable of educating and amusing the children, keeping the family budget, helping with the cooking and many other useful abilities" (Hoess cited in Rowlands 1978). However, while colleagues have told me of friends who were keen to use a micro to organize their hobby of choice, and financial software did sell, I find it difficult to believe that tasks such as household budgeting and recipe filing would have persuaded many prospective purchasers that the computer was indeed a useful piece of technology.[4] These were far from the "killer app."

This point is borne out in the discourse about penetration in computer magazine columns. While people continued to use first-generation, low-end micros well into the latter part of the 1980s, a "crisis" of low user numbers was being reported by mid-decade (recall that Ironmonger, Lloyd-Smith, and Soupourmas estimated that 6.6% of households owned a computer in 1985). In his August 1985 column in *Online: The Microbee Owner's Journal*, Eric Lindsay explicitly links the low penetration of microcomputers to the perceived uselessness of computers in the home:

> In home computers [in the UK], there were over five million in use, or an estimated 25% market penetration. . . . Micros just aren't going to sell the way TVs or toasters do. Less than one person in ten is likely to ever get a micro, unless it is included in an appliance. I personally believe that without fairly fierce advertising, and pressure on parents to buy micros for school children, the totals would have been even lower. Perhaps the razzle dazzle companies are finally running into their natural limits, and the micro market may be ripe for a return to purchases only when the micro has a use, and is evaluated on the basis of value for money. (Lindsay 1985, 40)

Lists of best-selling software published in January 1984 bear out Lindsay's points about usefulness and value for money. Considering these tables of sales by unit and dollar value by "four of Australia's leading software distributors" for the first few months of 1983 (see figure 2.1), we see that utilities such as word processors and database and financial software were purchased, but this differed significantly by computer brand. Imagineering helpfully divided their sales figures by system, and it is clear that while games dominated for the Vic-20, Commodore 64, and Atari computers, Apple sales were a mix of both games and utility software, while purchases for IBMs—with the exception of *Flight Simulator* and *Zork I*—were heavy on utilities. Clearly, in the early 1980s, there were (at least) two markets for microcomputers and software—a popular, low-end market and a business-oriented market. This is hardly surprising, given the difference in specifications between the IBM PC and the Vic-20, for instance. It was the difference between what it was possible to do with IBM's 16K of RAM (expandable to 256K) as opposed to the Vic-20's 5K of RAM (expandable to 32K). While games for the Apple II were highly regarded, the price of the machine tended to put it out of reach of most home users; thus, none of my informants coded for the Apple II.

The difference between purchase and actual use also needs to be borne in mind here, as does the possibility that the typing, word processing, and financial utilities for the low-end micros (Vic-20 and Commodore 64) came

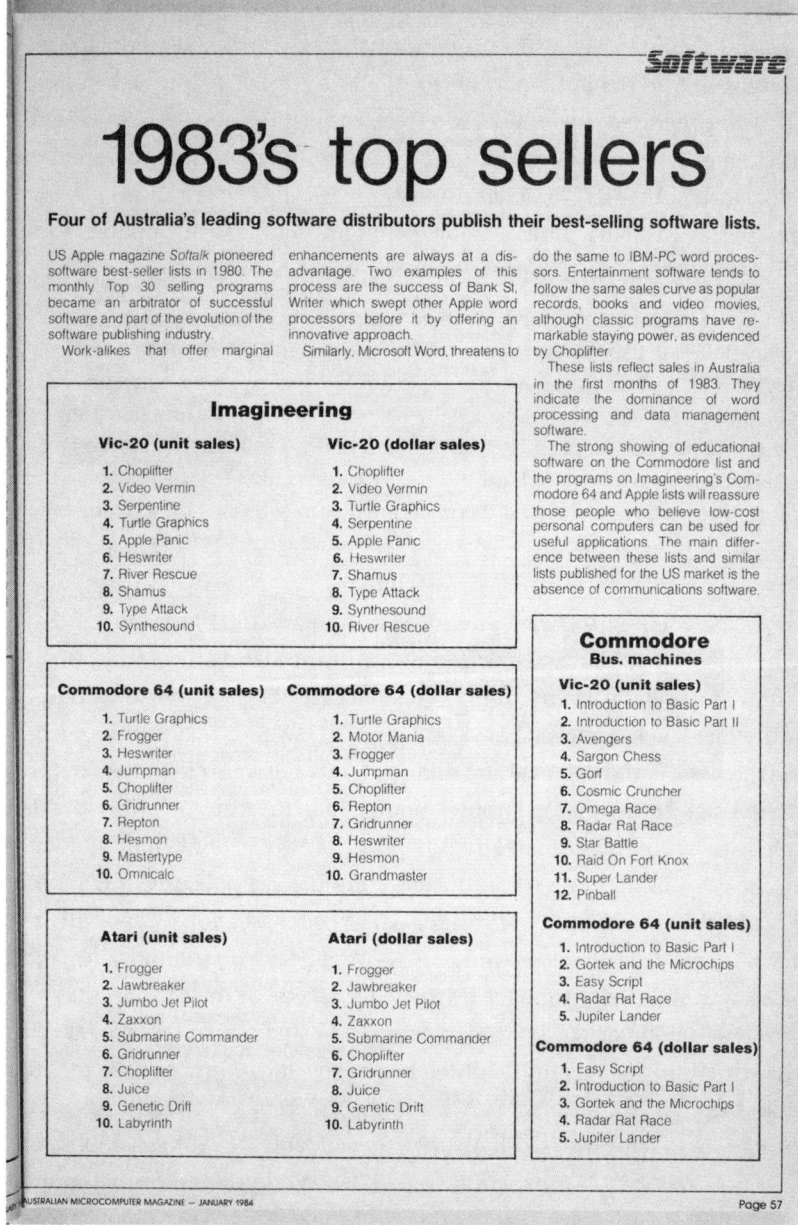

Figure 2.1
Best-seller lists, *Australian Microcomputer Magazine*, vol. 1, no. 10, January 1984, pp. 57–58. © Computerworld. Collection of the State Library of NSW.

Discourses about Microcomputers

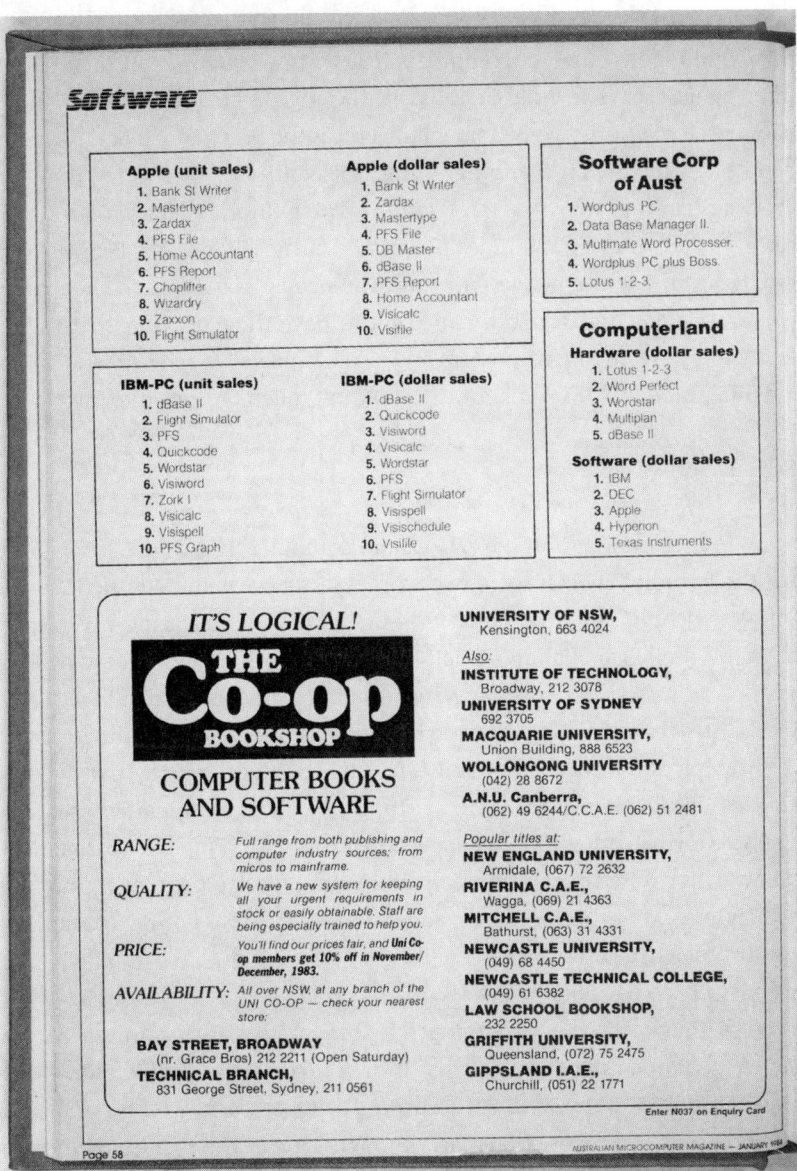

Figure 2.1
(continued)

bundled with the machines, a common practice at the time, as a 1988 ad for the Commodore range of computers clearly shows (figure 2.2).[5] One of my informants, Darryl Reynolds, recalls that he wrote "a range of electronic home packages for running your home, budget, cashbook, etc.," though "whether people actually used it is another thing." He continues, "The other thing, for every system that we wrote for, I wrote a typing tutor. I had the typing tutor early on and you've got no idea how many of those I sold. . . . People would buy it, again like the budget I'm sure. The people marketing to the retailers would go along and say, 'oh you've got to have these on your shelves, you know to justify buying the system and also it's practical.' So every system had a typing tutor and, do you know what, I can't touch type even today!"

Serious Software People

Neil Birss, cofounder of the New Zealand computer magazine *Bits and Bytes*, felt the tension between these two market sectors—what he termed "serious software people" and "home users"—firsthand. The magazine was pitched at home users, gamers, and small-business users, but these different audiences didn't always sit easily together, as Birss explained to me: "We hired a local Christchurch cartoonist to do the covers, but he was exceedingly savage in his depiction of accountants. I actually liked them but my partner, Paul Crooks, didn't. He was in charge of the ads; he was selling advertising. For example we had an edition where we tried to get accountants to advertise—because they were amongst the early users—but his depiction of accountants as hawks . . . well they wouldn't advertise again. So we got rid of him after two or three issues."

Birss observed that there was an "iron wall" between "serious" software people—that is, those who developed for mainframes and business—and the gamers, commenting that "a lot of the early computer people didn't want to get involved—they were frightened of being associated with a magazine that was involved with games. They thought that in the eyes of the business market that might taint theirs as a games machine." Such fears of diminished credibility on the part of business no doubt responded to and fed off the often negative media coverage of digital games—particularly in arcades of the period.

If there were doubts about microcomputers' domestic utility,[6] there was considerable ambivalence about their usefulness in the workplace. A number

Figure 2.2
Australian radio celebrity John Laws advertising Commodore computers. *Australian Commodore and Amiga Review*, vol. 5, no. 2, February 1988, p. 11.

of articles appeared questioning whether cheap micros would be of any use in business settings. In more-popular computer magazines, descriptive "case histories" would sometimes appear on a particular business's adoption of a microcomputer, as in Ric Richardson's series of articles in the *Australian Commodore Review* (Richardson 1985a, 1985b, 1986). These aimed to show that a Commodore 64 could be useful in a small business. Richardson concludes by admitting that his piece is "a conscious effort to get you, our enthusiastic games players and programming hobbyists, to use our familiar friend [the Commodore 64] for much more rewarding benefit IN THE WORKPLACE" (Richardson 1985a, 37).[7] However, the recurrent derisory comparison of micros to toys indicates that such low-end machines were seldom deemed to be sufficiently serious computers for business use, a theme that is clear in Frank Lee's review of four low-end micros in his article "Computing on the Cheap" for *Your Computer* magazine (Lee 1985). The elitism comes through loud and clear in the following blurb: "Lee dons his brown paper overcoat for a stroll down Cheapside Lane, hoping his Ritzy friends with PCs won't recognise him or the El Cheap Micro tucked beneath his arm" (Lee 1985).[8] Usefulness was thus equated with seriousness, and seriousness with computing power.

Reviewers in *Pacific Computer Weekly* (*PCW*) pronounced on "the line between the toys and real business systems," talking up the affordability and usefulness of the business microcomputer and "the ubiquitous CP/M operating system," but it is clear from the price that they are talking about high-end machines:

> Ten years ago, computer systems were restricted to large companies. Today, one can purchase a complete business microcomputer with software programs for less than $6,000. Such a price is comparable to the outlays a business might make for a copier, or several typewriters, and is considerably less than the salary of a qualified clerical person....
>
> Business microcomputer application programs of general use are the most readily available: word processing, payroll, accounts receivable, accounts payable and general ledger. More specific programs such as order entry, inventory control, doctor/dentist patient billing, professional time-cost accounting and accountant client write-up are offered, and will increase. (Anonymous reviewer 1983)

Like most of the periodicals written for the computer business, *PCW* showed little interest in cheaper micros; they were not part of its core business, at least not at that moment. But they didn't heap scorn on the smaller machines. The *PCW* article concluded that micros were getting more powerful, so it really came down to the software and whether it did what was required. This is a significant observation, as the low-end micros were often released before a range of software was available for them. As Philipson writes, "Without software computers are useless. The story of the computer industry is as much about programmers and the software they write as it is about the hardware" (Philipson 2004, 16). It was therefore entirely possible that an owner might not be able to use their computer as envisaged because of a lack of software—lack here meaning not just that they didn't yet own it but that it didn't yet exist. By contrast, what is evident in these business newspaper and magazine musings is the assumption that the tasks the computer would perform were already known; little scope was allowed for other new uses being discovered along the way. IBM's and Apple's pitches to the business market were premised on the automation of habitually performed business tasks, whereas the limited memory and processing power of low-end micros meant they were more constrained in what they could do. Such constraints also bred a creativity in imagining potential uses to which these micros could be put, as I will detail.

A Technology in Search of a Use

While some technologies' uses are implied by their function, I want to suggest that early microcomputers were effectively a technology in search of a use (or uses). Recall Eric Lindsay's very domestic toaster example cited earlier. The intended use of a toaster is obvious—indeed, it only does one thing—and, as long as you have electricity and some bread, it is immediately usable. The situation with early microcomputers was more complex. Given that some degree of coding was required to run one, a computer was not immediately usable by the uninitiated. Some saw this as a sign of uselessness, while others simply could not envisage how they might make use of one.

What Were Computers Actually Good For?

Another way for consumers to ask about the usefulness of early microcomputers was to ask: what will they do, what are they good for? I pose this question now to consider some of the actual uses that early microcomputers were put to by their owners. One thing that early microcomputers were definitely good for was playing games, and games have a special—though not unproblematic—place in this discourse of utility. In 1978, Rudi Hoess, the managing director of Electronic Concepts Pty Ltd (a computer shop), offered an answer to the twin questions that he said had been most prominent regarding computers: "What will it do?" and "Can it do something other than play games?" Hoess responded thus:

> If one analyses what computers do, the answer to the introductory questions are simple.
>
> For one, computers make logical decisions based on a pre-arranged (programmed) path taking into account variables supported by the user—and that is exactly what is happening when you play a game (computerised or with human partners)—the rules are set (programmed), while the inputs are based on player inputs.
>
> Indeed, the game is the most readily accepted way to make use of the new capabilities suddenly offered by new personal computers. (Hoess 1978)

Digital games were a new cultural form that was indigenous to the computer and took advantage of its abilities. Despite players first encountering interactivity in the arcade—where machines ranged from electromechanical, through transistor technology, to digital—games as a form were completely *at home* on the microcomputer.

There is good evidence that games were purchased widely by Australian micro owners. As already observed, games feature heavily in the *Australian Microcomputer Magazine*'s best-seller list of software for 1983, occupying all ten spots on Imagineering's list for Atari computers ("1983's Top Sellers" 1984). Such lists support Campbell-Kelly's claim that "games accounted for about 60 percent of home computer software sales"[9] (Campbell-Kelly 2003, 276), as well as later ABS research that found that games were used in 62.1% of Australian households where a computer was frequently used (Australian Bureau of Statistics 1994, 2). "To play games" was a common reason for purchasing a computer, even in the face of claims that it was really for "programming" or "educational" uses (Haddon 1988). Of course, not everyone thought that games qualified as either an example of computers' usefulness or a particularly good use for computers. However, gaming is clearly one use, and claiming that time spent playing games is time wasted is not the same as saying that computers are useless, though playing games might be seen as a nonprofitable or antiproductive use.

Despite the frequently expressed view that playing games was a waste of time, people learned a lot about computers in the course of playing games. Indeed, early games hold an important place as one of the great computer familiarizers, introducing people to the then new technology. Because games were pleasurable to play, such introductions could be fun (Butterfield 1978) and nonthreatening (Mendham 1986). In drawing attention to the element of fun, Mendham and Butterfield each observe a puritan ethic at work around computing, with Butterfield writing, "There seems to be an underlying feeling that there's something wrong with enjoying yourself" (Butterfield, 104). The fun of gameplay is sublimated in a 1988 advertisement for the Commodore 64 (figure 2.2), as purchasing a Commodore machine is pitched in terms of the familiar adage of "keeping up": John Laws—a well-known Australian radio personality—intoned that a Commodore was one of the best ways to "introduce your family to the world of computers." With their limited specs, games were one of the things that these low-end micros *were* good for.

As a computer retailer, Hoess was clearly hoping that some of the visitors having fun with the games in his shop would translate this into an abiding interest in computers. Interviewed by journalist Grant Rowlands, Hoess made it clear that people were welcome to come and use the computers on display:

> Mr. Hoess believes that by letting teenagers play games on the computers he is doing more than just entertaining them.
>
> "The children will begin to learn what a computer is all about and how rewarding it can be," he says.
>
> "They will see how easy a computer is to operate and program and want to learn more. . . . Today's fan is tomorrow's hobbyist and quite possibly the day after tomorrow's businessman." (Hoess cited in Rowlands 1978)

In contrast with the Commodore ad just mentioned, Hoess seems to have been articulating a more nuanced link, whereby playing games could be a path to more in-depth computer knowledge.

Programming

While games' strong showing on software best-seller lists demonstrates that users in 1983 were certainly purchasing games, the connection between early microcomputers and games went far deeper than this. Users typically learned some simple programming while playing games, especially in the early days. Even purchased software needed to be loaded from a tape or disk, as Mark Sibly and Simon Armstrong explained in response to my asking how they learned to code:

Sibly: It was much easier in those days, because you turned on a computer and you basically had to program. That was all you could really do. So, the first thing you had to do was command to load a program and then run it.

Armstrong: And you've learnt two commands, "load" and "run," and you go from there.

Many of the games that users played—particularly on low-end machines—did not come on tape or disk. Instead, they were distributed as source code listings, which users laboriously entered into the computer. Therefore, users virtually had to learn to program. As Katharine Neil put it, "You couldn't really do much with computers back then unless you learnt a bit of code. You'd do really dumb, primitive things, but . . . in those days, people bought games and they'd play games, but the coolest thing was to write stuff yourself. In those days, you bought a computer and you bought a book on how to program it, and there was only one way you could do it! And if you didn't do it, then what was the point of having a computer? Because it didn't do anything, it didn't do anything for you."

While programmability is of course not limited to computers, programming was a use that was indigenous to the computer—in the sense of *belonging*—and a whole genre of "how-to" programming guides appeared in the late 1970s and early 1980s to help users acquire programming skills. Some were published locally, while others were imported, such as *More TRS-80 BASIC*, coauthored by Bob Albrecht—one of the founders of the People's Computer Company in the San Francisco Bay Area—for the TRS-80 (Albrecht, Inman, and Zamora 1981). Owing to legal deposit legislation, the State Library of New South Wales (where I conducted this research on code books) holds an extensive collection of locally published titles, giving an insight into the computers people were buying and programming on. The library holds titles for learning BASIC for the Dick Smith System 80, the Exidy Sorcerer, the Microbee, the Apple, the BBC, the Vic-20, the TRS-80, and the Commodore 64, as well as a range of generic BASIC computer books, and manuals from education providers. A title published in the UK addressed coding for the ZX Spectrum.

Coding books are valuable for gaining a sense of how learning to program was presented. Because of the number of exercises that involve typing in code, it was clearly expected that the reader would be sitting in front of a computer as they worked their way through such books.[10] Books teaching how to code typically promoted the ease of programming a computer, providing self-study resources and the encouraging message that anyone could do it. While there were also books focusing on the craft of programming and some that addressed a particular professional or vocational sector (such as programming for accounting; see Scorgie and Magnus 1984), most were written for the home user.[11] Many began with how to turn on the computer, while others assumed "only that you've read enough of your User's Guide to be able to turn your computer on and that you're familiar with the functions of the various keys" (Oliver 1984, 5). Guides typically aimed to avoid "computerese" (Chalmers 1985, preface) and prided themselves on using "simple, logical, nononsense" language and a "light conversational style" (Wolff 1982, preface). Monro, for instance, confessed his long-held desire "to write a real computer book for real people" (Monro 1982, preface). The philosophy espoused in most of the titles is one of freedom, fun, and self-correcting learning.

Some code books focused on a particular system, while others attempted to cover the most popular brands of computers, such as Roz Ault's (1983) book *BASIC Programming for Kids*, which gives instructions for the most common brands for homes and schools in the US, which at the time were Apple, Atari, Commodore, Texas Instruments, Timex Sinclair, and Tandy Radio Shack. Different microcomputer brands used subtly different "dialects" of BASIC. The first chapter in Ault's book goes through the special keys on the keyboards of each of these computers: "If you have one of these computers, you'll find specific instructions for your machine in this book. But you should be able to use this book with just about any personal computer. The examples will work pretty much the same way on all types of machines. Just keep your manual handy in case you get stumped about which keys to press to make something happen" (Ault 1983, 1).

Compatibility between different micros was not assured, and their differences drove the production of games in important ways, as we will see. For instance, Street advised that "much is to be gained by adapting programs from one dialect of BASIC to another"; he considered such translation "a time consuming, but very instructive practice" (Street 1983, viii).

Quite a large number of code books explicitly addressed children, such as the twenty-eight books being given away in the *Sun Herald* (Sydney) newspaper's "Super Scene" children's liftout in 1985. The range included *Fantastic Game Books* for nine-year-olds and above and *Space Adventure Books* for eleven-to-fourteen-year-olds (each with a version for the Vic-20 and C64) (*Sun Herald* 1985). Throughout the literature, it is common to find references to children's abilities with computers, their lack of fear, and the importance of having them learn about computers, given they'd have to work and live in a world in which it was assumed that computers would become increasingly important (Ault 1983).[12]

Growing up in remote Westport, an isolated region of New Zealand, with little contact with others who were programming, Fiona Beals reported finding the Usborne range of books useful. With titles that include Craig Kubey's *The Winners' Book of Video Games* (1982), Ian Graham's *Usborne Guide to Computer and Video Games* (1982), and James I. Clark's *A Look inside Video Games* (1985), Usborne's titles appear to have prioritized the dissemination of knowledge and programs that could be used immediately, principally computer game code. Despite writing her own games and other programs from an early age, Beals recalls not knowing that "all those things were actually a computer language" until she learned BASIC in high school and realized—in a sort of "strange anamnesis" (de Certeau, Giard, and Mayol 1998, 153)—that she already knew it.

Anxiety at the idea of operating a computer was clearly anticipated, judging from both the topics discussed and the mode of address of many of the programming handbooks. The books sought to demystify computers, debunk various myths, and reassure users that they would not hurt or wreck the computer by pressing the wrong key (Inman et al. 1981; Micchia 1985). As Mendham writes, "Many films have shown computers exploding because of something typed in, and it is hard to persuade people that this simply does not happen" (Mendham 1986, 14). Humor is often deployed to combat misconceptions, as in *The Bewildered Parent's Guide to Computer Programming*, a Pitman Press title written by a high school student, Shane Micchia, who writes, "I *know* about bewildered parents because I've had first-hand experience. After all, my own parents became bewildered soon after *I* got my hands on a computer. I could feel their apprehension as I started to talk about loops, GOSUBs, ROMs and RAMs over breakfast. I began to feel sorry for them" (Micchia 1985, 1).

Alongside lots of cartoons—to begin with, of relaxed-looking teenagers at the computer with worried parents looking on, scratching their heads—the teenage Micchia devotes his first chapter to "removing some misconceptions," such as the belief that computers are smart (computers are "dumb"—Micchia 1985, 3) and good at math.

Beyond reassuring readers, many titles have the further goal of building confidence. For example, Ault writes, "It's easy—ok, so the computer doesn't speak English, you need to learn its language, but computers can be easy and fun for just about everyone" (Ault 1983, iii). The claim that anyone could learn to code was rife. It was made often, usually in relation to BASIC. It is, however, also extended to machine code in a sixteen-page free pamphlet I discovered in the Auckland City Library Ephemera collection, titled "Machine Code Made Easy":

> Add sparkle to your arcade action games, make your lasers zap aliens that no laser has zapped before, and utilise your utilities to their full potential. Make your aliens glide smoothly across your screen and make even the Spectrum appear to have *real* sound. Extend your machine's Basic to do all the things you could ever want it to do. Oh, the joys of machine code!
>
> "Ah, but . . . ," we hear you say. "Isn't machine code that difficult subject I read about in my manual?" Well, yes it is, but it isn't really all *that* hard, and with this series of cards, anyone will be able to get to grips with the subject.
>
> Every popular home computer except the Dragon is covered fully in these cards, and very soon you'll find yourself writing programs to rival anything you can buy. Well, fairly soon. ("Machine Code Made Easy" n.d., emphasis in original)

Experimentation

Complementing the ubiquity of the claim that programming was easy are the also common assertions that coding was creative and experimental. John and Judy Deane, for instance, wrote that "computer programming is not an arcane art or even a mysterious science. It is a creative HUMAN activity" (Deane and Deane 1980, iv, emphasis in original). Hoess compares coding to pottery, asserting that "the idea . . . that computer activities are uncreative as compared, say with rotating clay against your fingers until it becomes a pot—this is categorically false." He continues,

> "Computers involve imagination and creation of the highest level."
>
> "Computers are an involvement you can really get into, regardless of your trip or karma."
>
> "They are tools, they are glorious abstractions: so if you like mental creation, toy trains, or abstractions, computers are for you." (Hoess cited in Rowlands 1978)[13]

The authors of a 1981 book on the TRS-80 wrote with startling and penetrating clarity of the way that new uses would be invented for the microcomputer as new users envisaged them:

> Today's microcomputers, such as the TRS-80, have created an opportunity for nearly everyone to own, use, and master the "mysteries" of a small computer. The

small computer is finding its way into the home, the school, and the small business. Thousands of programs are being developed for these machines to perform a wide variety of educational, recreational, and business-related tasks.

With the steady increase of programs and computer users, the number of new applications for these powerful tools will continue to grow. *New users bring new interests and avocations that lead to different problems to be solved.* Everyone benefits as new solutions are discovered and shared, opening up new areas to apply the tool. The cycle feeds on itself, and everyone has an opportunity to be an inventor and creator in this rapidly growing field. (Albrecht, Inman, and Zamora 1981, 2, my emphasis)

Users who developed new programs were innovators in two ways: first, in the mundane sense whereby something had been created where previously there had been nothing. But it is the second sense that most interests me in this book, whereby programs were created by people who brought their interests and "avocations" to the task. This is consistent with Philipson's claim cited earlier that "many people who bought expensive and underpowered PCs wondered what to do with them." I assume the comment is intended to be derogatory, but it can also be taken literally: "wondering what to do" with something leads to experimentation and to new uses being found. (Experimentation extended beyond software to hardware also, a subject I discuss in chapter 5.) Such wondering on the part of hobbyists enabled them to see opportunities for developing new programs that would be useful in a field of their own interest or expertise.

Hobbyist programmers approached the computer as an experimental opportunity, a chance to improvise and develop programs of their own. There was a remarkable curiosity and a desire to find out for oneself what microcomputers could do. Many hobbyists began by writing games. This is how the history was narrated at the time,[14] as seen in Sharon France's potted history of the Microbee computer from 1985: "In the beginning the first pieces of software to appear on the market for the Microbee could be broadly classed under the heading 'Games.' Many of the titles were written by enthusiasts and were submitted to Applied Technology for appraisal, tidying up and an eventual view to marketing" (France 1985).

France's description of enthusiasts writing game titles for the Microbee and sending them in to Applied Technology is precisely the story of two of my homebrew game developer informants, Vaughan Clarkson and John Passfield. Clarkson and Passfield sent games off to Microbee's software publishing arm, Honeysoft, and these were subsequently published on tape and advertised to the installed base of Microbee users.

Magazines

Apart from tape or disk, the other main method of distributing programs—including users' early software experiments—was paper based. Many magazines sprang up to cater to early computer users, and these contained listings for games and other programs. Programs were a key reason for buying a computer magazine in the 1980s. Renato Degiovani's reflections on his period editing Brazilian magazine *Micro Sistemas* captures something of the energy and curiosity surrounding type-in software programs in magazines: "With issues that were quickly sold out in newsstands, it was eagerly read by users of personal computers, who looked for anything that would do something else with their computers and that they had not tried yet. The lists published were typed all night long, because this was the only fast way to obtain a computer program" (Degiovani 2003).

Users in Australia and New Zealand enjoyed some imported computer magazines, such as *Zzap!64* and *Computers and Video Games*, but there were also a range of popular local computer magazine titles (*Online: The Microbee Owner's Journal, Your Computer, Australian Commodore Review, Australian Apple Review, Sega Computer, Computer Input, Bits and Bytes*) as well as electronics magazines with sections dedicated to computing (*Electronics Australia, Electronics Today International*).[15] Magazines are important primary sources, and the local titles give a strong sense of the local culture of early amateur programming and computer use. A discourse of participation was strongly articulated and espoused in magazines, as my discussion of the New Zealand magazine *Sega Computer* will demonstrate. *Sega Computer* was set up by Grandstand Electronics, the distributor of the Sega SC3000 microcomputer, to help them compete against Commodore. Being a North American computer, the Commodore had plenty of English-speaking software available. By contrast, the Japanese Sega SC3000 was distributed in very few other English-speaking countries so it lacked a software base. Setting up a magazine and a "club" for users were shrewd marketing strategies employed to generate interest in the Sega line of products and ultimately enable it to compete with Commodore—then the top-selling computer worldwide—in New Zealand. A typed invitation from Grandstand sales manager Philip Kenyon to join the "club" expounded the benefits users would derive in return for their $39.95 subscription fee: owners were to receive the first six issues of "New Zealand's first dedicated computer club magazine," plus two free programs on cassette (Kenyon n.d.).

Between 1984 and 1986, Grandstand put out around ten issues of the magazine. By November 1984, they were touting *Sega Computer* as the "Official Sega User Club Magazine."[16] The magazine's contents included type-in programs, reviews, letters, advertisements, and technical articles on topics such as random numbers, program languages, and machine code. Grandstand also promised to publish the meeting dates and venues of anyone wishing to start a local user group. This decision probably partly accounted for the volume of Sega user groups that sprang up across the length and breadth of New Zealand: the August 1985 issue shows fourteen Sega user groups from Timaru to Tauranga.

While Grandstand may have begun the magazine to serve its commercial needs, the publication quite quickly took on its own significance in terms of the boost it gave to the local community of computer owners. Users were invited—indeed expected—to contribute to the magazine, and it is clear that they did. Employing the language of clubs and membership, the magazine strongly encouraged a participatory ethic. Philip Kenyon's invitation in the first issue is representative:

> This is your magazine, so we welcome your suggestions on what you want in it (no suggestions on where to put it if you please). Your support is needed to maintain the quality of the magazine, so come on all you budding superbrains—start sending in those programs and letters—all printed on your shining new Sega Plotter Printer, of course.
> I look forward to hearing from you in the near future. (Kenyon 1984)

The editors were particularly keen that readers contribute their programs, which they did. As well as publication, the best submission in each issue would typically earn its author a small prize. Having a program published in a magazine provided recognition for a user's talents. Auckland teenager John Perry recalls that people were "pretty impressed" when, at the age of thirteen, Grandstand opted to publish his game *City Lander* on tape. This was not the first time he'd "made it": Perry previously had a program he'd written, *Harbour*, published in *Computer Input*. But the second time around, Grandstand arranged a television spot on TVNZ's *Top Half* as publicity for his achievements (see figure 2.3). Setting aside Grandstand's clear self-interest in such instances, the company also spoke strongly in favor of locally written software, written by young people—literally titling one piece "Sega's Young Programmers: Today New Zealand, Tomorrow the World" (Anonymous 1984)—and in the time that they marketed the Sega SC3000 system,

Figure 2.3
John Perry on TVNZ's *Top Half* with his game *City Lander*. "Computer Kid" segment, *Top Half*, June 1984, Television New Zealand, via Getty Images. (See plate 1.)

they published hundreds such software titles for it, making them a significant publisher (Wheeler and Davidson 2008).[17] Grandstand's support evidently encouraged the young community of hobbyist programmers. Readers' appreciation—and criticisms—appear often in the letters pages. Sometimes, compliments are simply expressed in the form "P.S. Great magazine," but the short letter from F. K. Maynard of Wellington sounds a common theme: "The article 'Program Dissection' in the September issue of Sega Computer is the most helpful we have read. We hope there will be many more such articles to follow. We look forward to receiving further issues of this magazine" (Maynard 1985).

Maynard would receive another year's worth of issues before Grandstand passed the magazine on when it became the Amstrad distributor. After that, *Sega Computer* (and the Sega club and club support) would be passed on to a series of others to edit and manage, first to Glenys Millington of Sega Software Support and then—when she in turn passed it on—to Poseidon Software. This arrangement again did not last long: in a letter to magazine subscribers, Poseidon claimed to have a "large amount of evidence that the magazine is being photocopied in numbers and resold." Together with low subscription numbers and in anticipation of the launch of the Sega Master System, this led to their outsourcing publication to Michael J. Hadrup (Crawford n.d.). Hadrup took this on while he was a senior student in 1987, putting out roughly five issues—some of which were doubles—before finally closing it down in 1988 (Swalwell 2010).

Magazines such as *Sega Computer* were undoubtedly a key support for a nascent community and culture of computer enthusiasts, offering participatory forums for discussions about programming and technical matters, recognition for achievements (programs published and game high score columns), and contacts for user groups. Anecdotally I have heard that Australian Sega owners were jealous of the support that seemed available to their New Zealand counterparts but was apparently not forthcoming from the local distributor, John Sands.

User Groups

User groups involved getting together with like-minded others to share know-how and resources in the tradition of other early technical hobbyist clubs (e.g., ham radio, meccano, the nineteenth-century model construction system).[18] Typically run fairly informally, the groups provided a social context for computing. They were one of the first times that people got together socially around, and because of, personal computers. On their own, the lists of user groups that were publicized in magazines don't convey the level of enthusiasm these groups exuded in their early phases, when, as Neil Birss put it, "it's driven by the folk themselves" who have that "let's hire a hall somewhere" momentum.

Writing on user groups for the Dick Smith VZ computer, Bob Kitch sees the shift from tinkering around with a computer to forming a user group as a natural development: "Users and owners . . . naturally tended to band together, to chew over mutual interests and problems. . . . [that and the poor support they received from Dick Smith Electronics, according to Kitch] These 'jam' sessions were most often held over the phone, but have you ever tried to satisfactorily discuss a software problem over the phone? The next stage was to organize a meeting of interested enthusiasts, usually on a weekend, in someone's home or at a conveniently located hall. And so began a VZ USER GROUP" (Kitch n.d.).

When asked to name his best memory of the Sega SC3000 era, Robert B. Brian (author of the *Sega Programming Manual* and numerous Grandstand programs) nominated a user group meeting: "One of the very first meetings of the Wellington Sega User Club was held in my flat in Adelaide Road. I think meeting other users and sharing ideas, Software, FAQ, etc., was great" (Wheeler 2008b).

Other Homebrewed Software

While games may have been among the earliest homebrew programs to appear, the range of uses hobbyists could see for the computer didn't end there. With a limited range of software in the early days, innovative non-game software titles also began to appear. Computing magazines often mentioned them in articles. For instance, several intriguing creations by Robert Bowden (US) were editorialized in the *Australian Commodore Review*. *Period* was claimed to be "the first software program that can be used as a guide to determine when a woman is most likely to conceive." *Planning Tanning* was a program from the days before skin cancer awareness, where the user "tell[s] the computer the month, time of day, your skin type, sky condition and the type of tanning lotion you're using. Then the computer would tell you an exact tan time for each side of your body." Finally, *For Ectomorphs Only* was "a program to help you gain weight and increase your strength and muscle mass. [It] will also help skinny people pick out the most flattering clothes by demonstrating how colour, style, and pattern affect appearance" ("Birth Control via Your C64" 1988, 4). While some of these program ideas might seem bizarre, they highlight the argument I am making that hobbyist programmers developed software for which *they* could see a use. These were everyday problems or issues that users began to think about in terms of computation. At times, a shared commonsense view of what programs people were *likely* to find useful is also evident in magazines. For instance, in passing on a request from a Swedish Microbee user for astrology software, the editor of *Online* notes that a number of people expressed an interest in this, so surely *someone* must have written a program along these lines or would like to do so (Anonymous editor 1984).

Other examples of software written during this period that addresses everyday vernacular problems include the output of Armidale entity ArComPro, which developed an extensive list of software programs, including *Auction Lots, Bar File, Radio Operator's Logbook, Beef Stud File, Pony Jamboree,*

Showjump, Genealogy, Warranty Recorder, Weight Recorder, Sharemarket, and *Squash Controller.* The State Library of New South Wales holds documentation for two of their games, *Olympic Gold* and *Quizmaster. Compu-B,* by John Schellens and James Roe, was a horse racing analyzer distributed by Dreamcards, a business run by Lindsay R. Ford, a lawyer from Victoria. In describing the program, Ford writes that it is "probably the only microbee program that can truly claim to be capable of paying for itself" (Ford 1985). Ford's own programs included *Psychotec*, a "computer psychiatrist" program with artificial intelligence, and *Merlin*, an adventure game (Ford 1984).

Educational software was another field in which users began experimenting with creating programs for which they could see a need. Dean Hodgson was a teacher in Port Pirie, in rural South Australia. After buying a Tandy Model I in 1979, he taught himself BASIC[19] and started writing software for which he could see a use in the classroom. "I researched Computer Assisted Instruction (CAI) and made sure my programs were based on those learning principles," he explained. A few years later, his school bought two Tandy Color Computers, for which he wrote a few dozen programs for the entire school to use, including a school library catalog and several learning games that Tandy Australia later bought and distributed: *Maths Invaders, Spelling,* and *Cordial Stall* (Wheeler 2008a).

Anxieties about the perceived usefulness (or not) of early microcomputers were a recurrent feature of 1980s discourse on computers in the home, and it seems likely that this anxiety affected their uptake. Hobbyists seldom had any issues, being already convinced of the computer's "fiendish fascination" (Hoess cited in Rowlands 1978). Beyond the hobbyist realm, opinions were split as to whether early micros were not useful (useless) or whether they were effectively useless in their (then) current form, because of a lack of software, for instance. I have suggested that this was partly because none of the claimed early uses adequately caught the public's imagination. Writing and playing games was (and is) a use for which computers were ideally suited, but it was not a use that satisfied the criterion of usefulness for all. Hobbyists and others invented new uses for computers by experimenting. The software titles such hobbyists wrote were instantiations of their ideas of how the computer could be put to use. Many of these were niche applications, however, and those who had already decided that computers were not for them were not listening. It seemed to take until the release of applications with widespread appeal (such as telecomputing, music making, and desktop publishing) in the latter part of the 1980s for perceptions to start to turn around and for nontechnical users to get interested in what computers could do for them. Very practical applications, such as printing personalized stationary, seemed to strike a chord with the wider market. In chapter 4, I will return to this question of usefulness and consider the factors that helped to change perceptions of computers in the latter part of the 1980s to the point where nonhobbyist members of the population began to see computers as something that could be quite useful, if not yet something that they needed. Chapter 3 moves on from the discourses that shaped the micro's reception to actual accounts by homebrew creators of their micro use, the games they made, and how they understood their activities at the time.

3 Micro Users as Makers

In his book *Electric Dreams: Computers in American Culture*, Ted Friedman acknowledges the central importance of games to popular computing, yet games are largely treated as a consumer product to be played on personal computers. I find this curious given that Friedman profiles the introduction of the Altair computer and the hobbyist culture that surrounded it (Friedman 2005, 92–101). The attention to Altair hobbyist culture seems to largely serve a narrative about the first personal computer and how a nascent "privatized concept of computing" then developed, paving the way for a commodification of computing that Friedman suggests was "quintessentially American in its individualism and consumerism" (99). Friedman's next chapter extends this theme by focusing on the marketing of the Apple Macintosh and the "firm establishment [of the computer] as a mass-produced consumer item in the mid-1980s" (102). The Altair references notwithstanding, Friedman's account is largely a history of the computer as a consumer good, which has been the dominant approach.[1] But the computer as a consumer good is only part of the story. Computers—both in the early years and now—are more than just consumer goods. As my informants' accounts in this chapter will demonstrate, computers were something they experimented and created with. In the 1980s, games were not just something one played on a microcomputer but also something one made.

As foreshadowed in chapter 1, the work of French historian Michel de Certeau is particularly helpful for understanding consumption as a form of production and for encouraging attention to what consumers actually *make* and *do* with products. In this chapter, I first reconstruct some of the known facts around the publication of de Certeau's key works on these topics and reflect on their Anglophone reception. I then offer an account of the theoretical framework underpinning the studies of de Certeau and his collaborators Luce Giard and Pierre Mayol. From this general account, I move to deploy specific ideas from their scholarship to analyze aspects of 1980s microcomputing and homebrew development practice. In particular, I focus on the local orientation of practice, and cooking as an analogy for users' experimentation, to offer a detailed excavation of homebrew practice.

User Making

Though de Certeau is best known in English-speaking academia for his book *The Practice of Everyday Life*, he was a religious scholar, a Jesuit historian. He wrote a masterful account of the demonological neuroses that centered on the town of Loudon in the seventeenth century during a case of "possession" and trial for sorcery. Natalie Zemon Davis describes de Certeau's history of the possessed women of Loudun as "a pioneering ethnography of human relations and spiritual practices in the seventeenth century" (Zemon Davis 2008, 59). De Certeau, Giard writes, was an "anticonformist," "interested in the semiotic or psychoanalytic reading of situations and texts" (Giard 1998a, xiv) and "one of the rare historians [of his generation] eager for new methods [and] ready to venture into them" (xix). According to her, he was "feared for his demanding and lucid criticism of the epistemology that silently governs the historical profession" (xiv). De Certeau's preparedness to venture outside orthodox historical method is still refreshing and no doubt one of the features that made his work interesting to those scholars in the 1980s who were embracing interdisciplinary inquiry and casting around for new methods, as some forms of the cultural studies project did, for example.[2]

While *The Practice of Everyday Life* is a well-known text, the background to this book is less well known. It appeared in English in 1984, translated by Steven Rendall; however, the book had its origins in a three-year research project commissioned by the French Department of Research at the State Office for Cultural Affairs. Conducted between 1974 and 1978, the final writeup of the research was submitted in 1979 (Giard 1998a, xviii) and published as *L'invention du quotidian*. Significantly, Giard locates the origin of the study's questions in the events of May 1968, particularly the "perspective reversal" that "displac[es] the attention from the supposed passive consumption of received products to anonymous creation, born of the unconventional practice of these products' use" (xvii).

In the General Introduction to *The Practice of Everyday Life*, de Certeau demonstrates this perspective reversal by making a distinction between production and "the uses that are made" of products by "users who are not producers." The idea is akin to what we might think of as *consumption* but lacks the pejorative connotations that term often has in consumer capitalism. While this formulation—consumption as a form of production—is admittedly paradoxical, it is useful in that it facilitates inquiry into practices often passed over because they are not immediately obvious. As de Certeau writes, "The 'making' in question is a production, a *poiesis*—but a hidden one, because it is scattered over areas defined and occupied by systems of 'production' . . . and because the steadily increasing expansion of these systems no longer leaves 'consumers' any *place* in which they can indicate what they *make* or *do* with the products of these systems" (de Certeau 1984, xii, italics in original).

To briefly recap the gloss I gave in chapter 1, the verb *faire* in the French subtitle of the book, *Arts de faire* (literally, the arts of making or doing), clearly captures the consumption as production dynamic and more effectively conveys the sense of an active making than does the English term *use*, which tends to imply functionality or instrumentality. This term *faire* was employed by various computer fairs of the era, including the famed West Coast Computer Faire. The contemporary Maker Faires are clearly positioning themselves as the successors to this tradition.[3] The nuance inherent in this perspective reversal is easily lost, particularly when commentators attend to mass popular culture and its reception by audiences. We see this in John Fiske's appropriation of de Certeau in his concept of the "art of making do," an ambivalent concept that besides highlighting the active role that audiences play in deriving meaning from media texts implies an acceptance of constraint, and a level of resignation and lament in that audiences must make do with the cultural products that are on offer (Marshall 2004, 9).[4]

Audiences create culture and make their own interpretations of texts, but people are also literally makers, a point that has sometimes been obscured in cultural and media studies' appropriation of de Certeau, which has largely focused on audience interpretation of texts, often from popular (or mass) culture. De Certeau and Giard draw a distinction between mass and everyday or ordinary culture, which helps to tease apart some of the different contexts in which making is deployed. They write, "Mass culture tends toward homogenization, the law of wide-scale production and distribution. . . . Ordinary culture hides a fundamental diversity of situations, interests, and contexts under the apparent repetition of objects that it uses. *Pluralization* is born from ordinary usage, from this immense reserve that the number and multiple of differences constitute" (de Certeau and Giard 1998a, 256).

The terms in which de Certeau describes consumption in the Introduction to volume 1 have always resonated deeply with me, but they have also seemed slightly at odds with the rest of his book. Most of the chapters feature a strong emphasis on language and reading, their relation to orality and a "scriptural economy," and enunciation. Indeed, Luce Giard calls reading a "central paradigm" (Giard 1998a, xxiii). I have always found it curious that this idea of consumption as a form of production didn't receive greater attention. Indeed, it is not until chapter 12, "Reading as Poaching"—the metaphor that Henry Jenkins will borrow for his seminal fan studies book *Textual Poachers* (1992)—that there is much focus on consumption at all. The ways that de Certeau has been taken up by scholars have only confirmed what a missed opportunity this is. His distinction between strategies and tactics is often invoked, as is his work on walking through the city, and the concern with escaping a rationalizing and panoptic gaze, ideas that bear "an obvious echo of the work of Michel Foucault" (Giard 1998a, xx). I have admired those studies that seem to have gestured toward productive consumption and interpreted *faire* not only metaphorically but also as approaching something more literal, as making. Nevertheless, I have always found that the deep meditations and case studies fleshing out de Certeau's insights about consumption as production seemed to be missing, and I've puzzled over why this was.

I think I now know why: de Certeau's book that appeared in English in 1984 comprised only the first volume of the two-volume study that was completed for the French government. The "missing" case studies were published in a second volume, *L'invention du quotidian, volume 2, Habiter, cuisine*. Though I had come across and read some of this volume shortly after its publication in English in 1998 via a colleague in food studies, I had largely forgotten about it until I began to plan this book.[5] The analyses in this title—conducted by Pierre Mayol and Luce Giard—were intended to resonate with the theoretical framework of volume 1. Mayol and Giard had participated in several collaborative "circles" de Certeau had organized since 1974. The case studies grew, respectively, out of Mayol's interest in practices of "dwelling" in a working-class neighborhood in "the provinces" of Lyon and Giard's realization at one

Micro Users as Makers 53

of their weekly research meetings "that women were strangely absent" from the project (Giard 1998a, xxiv–xxviii). However, despite the importance of the case studies in volume 2 to the research project as a whole—de Certeau himself called them "an undoubtedly more important facet than the explanation of ways of operating and modes of action in the first [volume]" (de Certeau, Giard, and Mayol 1998, 3)—volume 2 was apparently thought to be "too . . . French" (Giard 1998b, xlii) and so was not translated into English until 1998 (by Timothy J. Tomasik).

The timing of the translations explains much about the ways in which de Certeau and his colleagues' ideas have been received and adopted by Anglophone scholars. All the studies that I have read that apply de Certeau's ideas on consumption to media audiences, users, and fans have been undertaken through recourse to volume 1. Two decades after the translation of *The Practice of Everyday Life, volume 2, Living and Cooking* (1998), at a time when media ethnography is a widely used methodology, there remains much to be gained from revisiting and engaging with de Certeau and his collaborators' work on consumption, legitimation, and authorization, concepts that speak powerfully to the concerns of this study. Indeed, the microcomputer user and homebrew game developer constitute some of the best examples of the insight that users and consumers are makers and producers of ordinary culture.

Volume 2 of *The Practice of Everyday Life* comprises two studies intended to "illustrat[e], through the details of concrete cases, a common way of reading ordinary practices, of putting theoretical propositions to the test, of correcting or nuancing their assumptions, and of measuring their operativity and relevance" (Giard 1998b, xliii). In part I, Pierre Mayol presents "Living," an in-depth study of a family living in the Croix-Rousse neighborhood of Lyon and the relationships they have in the area, particularly with shopkeepers. In part II, Giard attends to the arts of "Doing-Cooking." Despite the very specific emphases of Mayol and Giard in their accounts of ordinary French people's everyday practices—of shopping, cooking, and urban life in the 1970s—and despite Giard's acknowledgment in 1994 that certain "practices . . . have already receded from us" (Giard 1998b, xl, xliv), many of the central themes of volume 2 are surprisingly relevant and even applicable to 1980s homebrew game development and computing in the home, as I will detail in a moment. First, however, I need to piece together a little more of my journey through the different source texts.

Having solved the mystery of the "missing" case studies, I expected that the two volumes of *L'invention du quotidian* would comprise the core theoretical material I would use from de Certeau. Ironically, despite the fact that *L'invention du quotidian* was finished in 1979, there is very little mention in either volume of computing,[6] even though some microcomputers had been available for some years by then. This omission bothered me, yet the relative lack of reference to home computing at this time is hardly surprising, given its location in private domestic space.[7] (And Minitel—the peculiarly French digital terminal for connecting over the telephone network—did not eventuate until a few years later; see Mailland and Driscoll 2017, 9–13.) Moreover, I reasoned—with Giard—that it was not their task to compile an "'encyclopedic description' of everyday life" (Giard 1998b, xxxvii), and from other comments in volume 2, it seemed that de Certeau and Giard *had* anticipated the uses that would be made of a range of newer technical media, including the computer. As they write in the concluding chapter, "A Practical Science of the Singular," "It is false to believe henceforth that electronic and computerized objects will do away with the activity of users. From the hi-fi stereo to the VCR, the diffusion of these devices multiplies uses and provokes the inventiveness of users, from the manipulatory jubilations of children faced with buttons, keyboards, and the remote control, to the extraordinary technical virtuosity of 'sound chasers' and other impassioned fans of hi-fi. People record fragments of programs, produce montages, and thus become producers of their own little 'cultural industry,' compilers and managers of a private library of visual and sound archives" (de Certeau, Giard, and Mayol 1998, 254).

It was while I was doing some "holiday reading" from de Certeau's oeuvre that I realized why the sudden mentions of computerized objects, VCRs, and "sound chasers" in the last citation seemed out of place. They originate not from the 1979 study but from a 1983 report de Certeau and Giard conducted for the French Ministry of Culture, *L'ordinaire de la communication*.[8] A close reading reveals that the two reports are related projects. The piece titled "A Practical Science of the Singular" (de Certeau and Giard 1998a) literally connects them: originating in the 1983 report, it was included as a final chapter in the second edition of volume 2.[9] While the thick, very concrete descriptions of the earlier ethnographic case studies are replaced with more abstract language in the 1983 report, it nonetheless continues a number of the earlier project's concerns, as well as introducing some new ones and some different terminology. Parts of the report appear in the posthumously published title *The Capture of Speech and Other Political Writings*, which collects de Certeau's work. Giard edited and wrote an introduction to this book, as well as coauthoring sections (de Certeau 1997). *Orality, operativity*, and *the ordinary* are key terms, with many mentions made of "ordinary culture," "ordinary practices," and "everyday savoir faire" (de Certeau and Giard 1997, 101). These form a counterpoint to the "scriptural economy" that receives so much attention in volume 1 yet is referred to by de Certeau and Giard—in a segment taken from the 1983 report—as being (only) one half of culture: "A politics of communication cannot be based on forgetfulness or scorn for one of the two halves of culture: that of writing, legitimated production, and the scientific discourse of knowledge, and that of orality, ordinary practices, and everyday savoir faire. Both need to be developed, assisted, and encouraged, for each depends on the other, and each enriches the other through its rigor and invention" (de Certeau and Giard 1997, 105).

My analysis and appropriation of concepts for the present title ranges across these three texts: the two volumes of *The Practice of Everyday Life* and those chapters of *L'ordinaire de la communication* that have been translated.[10] I read these together, allowing Mayol and Giard's case studies to speak to the theoretical foundations de Certeau lays out in volume 1 and concepts from the 1983 report by de Certeau and Giard to speak to and enrich ideas outlined in the earlier commission.

One goal of the later report is clearly to reframe culture as something that is practiced rather than inhering either in material products—"Culture is judged by its operations, not by the possession of products" (de Certeau and Giard 1998a, 254)—or "what is most valued by official representation" (251).[11] Also central to the analysis are networks—both those "by which people are bound by passion or shared convictions [including] amateur practitioners of astronomy, computer hacking, or music, nature lovers, [etc.]" (de Certeau and Giard 1997, 108) and "cable networks" (93)—and references (finally!) to "experiences with the 'Web' and E-mail" (110), the new technologies of tape recording, videotape, "coding," (113) and computer clubs, and various practices of repair (113) and improvement. De Certeau and Giard write that such networks are treated as illegitimate and are held in low esteem: "In [these networks] circulate elements of knowledge and know-how, information about economics, geography, or technology. These are the real networks of communication and pedagogy, even though an elitist and abstract conception of culture esteems them to be negligible.... Here is invented and practiced a way of refashioning the sociocultural environment, of appropriating its materials, and of making use of them for different goals" (114).

The authors clearly see an energy in practices that are deemed illegitimate by an "elitist conception of culture." Using the example of amateur writing workshops, slam poetry festivals, and "expression that retain[s] a ludic dimension"—in other words, vernacular—they argue that "work needs to be done to destroy the artificial barriers that official discourses of knowledge have erected between written and spoken language. It needs to be explained and shown how one is always nourished by the other" (127).

Microcomputing as Ordinary Practice

> We know poorly the types of operations at stake in ordinary practices, their registers and their combinations, because our instruments of analysis, modeling, and formalization were constructed for other objects and with other aims.
> —de Certeau and Giard, "A Practical Science of the Singular," 256

After de Certeau and Giard, I suggest that we still know little about what people *actually did* with microcomputers. Although some published local accounts from the period attempt to evaluate the social and labor significance of the computer, for instance, these tend to be more concerned with either the spectacular (unauthorized hacking into systems) or the feared potentials of the "computer revolution," such as job losses (Beardon 1985). While such accounts are useful for understanding the fears and anxieties attached to the computer, they are of little help in understanding actual user engagements. And, as I demonstrated in chapter 2, the promises of what computers would be useful for do not provide answers regarding what people actually did with computers. Clearly, an ethnography is required, but as we are now more than thirty years after the period when these early computers were being actively used, direct observation is not possible. In lieu of this, I have interviewed former users to inquire how an admittedly select group of consumers—all of them home coders, most of them homebrew game developers—used their computers. In doing so, I am heartened by Tom Conley's remarks on the "Janus-like solidarity of anthropology and history: the historian and the ethnographer share the same road" (Conley 1988, xv–xvi).

In the rest of this chapter, I invoke specific ideas from Mayol and Giard's studies, in conjunction with excerpts from interviews with my informants, to flesh out two aspects of 1980s microcomputing as ordinary practice: the very local orientation of informants' computing practices and cooking as an analogy for users' experimentation and creation.

Home and Local Environs

Mayol's study ponders the nature of a neighborhood, city dwellers' use of urban space, and the rhythms of life in the Croix-Rousse. While he covers many aspects of the housing and habits of those who dwell there (shopping, wine consumption, familial relations, leisure, etc.), it is Mayol's comments on home and the neighborhood that resonate with microcomputer users. Like my informants, the inhabitants Mayol studied largely operated at a neighborhood level (deemed a "privatization of public space" by making it their own) and in the domestic private space of the home. Home, Mayol writes, is one of the few places where "one can *do* what one wants" (de Certeau, Giard, and Mayol 1998, 11, emphasis in original). Production in the home or neighborhood is not subject to the imperatives of work: unlike the workplace, which is about "necessity," there can be a "gratuitousness" in the neighborhood—and, I would add, in one's home—where ends need not only be functional (12–13).

As the term implies, the homebrewing of games typically took place at home—if not one's own, then someone else's. Figure 2.3 shows a young John Perry on his Sega SC3000, located in his bedroom. While some were lucky enough to have a "computer room" as in figure 3.1, low-end micros seldom came with a monitor, so they needed to be plugged into a screen, which was often the family television set. Figure 3.2 shows a young Andrew Stephen crouched over his ZX81 for just such a reason. It seems to have been quite usual to set up the micro in a communal space such as the living room.

Beyond home itself, the local neighborhood was also significant, particularly for those without their own micros. As young boys, Andrew Stephen and Ross Symons found outlets for their curiosity about, and fascination with, computers and games in a range of neighborhood venues in New Zealand and Australia, respectively. The local pub (bar), the computer store, and the news agency featured prominently in their exposure to the world of programming.

Figure 3.1
"Muzboz" at home in the family computer room. Photo by Brenda Lorden. (See plate 2.)

Figure 3.2
Andrew Stephen at his ZX81 in 1982, at age ten or eleven. Photo by Robin Stephen. (See plate 3.)

Micro Users as Makers

Andrew Stephen's interest in computers started early. His father has told him that he had been interested in the *idea* of computers since he was very young:

> The thing that really got me into computers was seeing a ZX81 being displayed in the store. I was just absolutely fascinated by it and started poking at the keys, and seeing things appear on the TV. One of the sales people came up to me after I'd been in there a little while and gave me the manual. The manual for the ZX81 was actually a very well written introduction to BASIC programming. So it wasn't just a language reference manual, it talked you through from the very early concepts all the way through learning to program the computer. So for a couple of weeks I stood in front of a ZX81 in a store, and plugged in the example programs from the book, and started modifying them and seeing what happened. I was going down after school every day and having a go on it, and getting a bit further and having a play.

Stephen observes that his frequenting of the computer store is similar to the way that people nowadays go to the local electronics store and use the demo console units at lunchtime:

> There are still people who'll be standing there in front of the X Box display or the PlayStation display, and having a game. But the difference was, back then of course, because it was new, people didn't really . . . know [what to do with a computer]. You could go into a store and you knew what a kettle was, and you knew what to do with one. You didn't need to have a play with every kettle in the store to understand the concept. But I guess with computing, you did need to experience it to really have any idea what the things were. So every shop I remember had row upon row of all the different machines: the 1K ZX81, the Spectrum, a Commodore, and a VIC 20, and an Osborne 1, and whatever, all put out, and people would go in and just play with them. And a lot of the time there was no preloaded software or anything, so people were going in and standing in the shop writing programs. And I remember writing demo programs for stores as well, just when I went in. I'd plug away, write them a program, set it running. And none of them were ever saved, none of them were ever retained. As soon as they were turned off at the end of the day the program was gone. And I wasn't the only one. There were other people, I remember, going to the stores and doing exactly the same thing.

After writing programs in the computer store, Stephen's parents bought him his own ZX81, when he was age ten or eleven. Once he had his own computer, he "started doing some hardware modifications on" it, a topic I'll turn to in chapter 5.

A young Ross Symons had seen and conquered *Pong* in the local pub in Frankston, an outer suburb of Melbourne, around 1978. This started him wondering how such games could be made. The pub "had a system where if you won the game you stayed on the machine, and the next person paid you 20c and paid for the game." Symons was making what he thought was a fortune, sneaking past his uncles into the pub to play *Pong*. Little did he know that it was his uncles who were making the real money from his newfound skill, evidencing what de Certeau considers the "ruses of consumers" and "the intelligence and inventiveness of powerless people" (Giard 1998a, xx, xxii). "What was actually occurring," Symons explains, was that "my uncles knew that I was there, in fact they were waiting for me to come, and they'd be hustling people. You know, they'd say to people, 'You couldn't even beat that kid.'"

Before long, Symons started delivering papers for the local news agency, which freighted in computer magazines from the US:

> [The magazines] were around 10 bucks which, in the 70s, was a hell of a lot of money. I actually did the paper round not for the money in the paper round, because I was earning more playing *Pong* than from the paper round. Essentially I did the paper round . . . to have access to the magazines. I'd get to the paper round an hour early and just keep reading and reading and reading, without any ability ever to have a computer because it was just way out of our means. I think at the time there was a thing called the Exidy Sorcerer and a few other computers that you could get from the U.S. and they needed 110 volt conversion to 240. It was a lot of trouble. So essentially all I could do was theoretically program. I actually started notebooks of code that were theoretical programs that I'd run in my head.

Later, when a Radio Shack store opened in Frankston, Symons would use the computers on display there to program: "I'd go in there, use their computers, they'd ask me to do small things and I'd just do them, and that's how I learned to code . . . going to Radio Shack and using their computers because I didn't have one. Little did I know they were using me to sell it, kind of the same way my uncles had where they'd say, 'Hey, that kid can use it; anyone can.'"

Another informant, John White, made use of the computer at his father's workplace in the Psychology Department at Victoria University of Wellington. White recalls spending time with his brother Joseph[12] on the computer during the school holidays. White doesn't remember what type of computer it was, just that it "had a little green screen embedded in this big box" and was "connected up to a big switchboard which was a manual interface for setting up experiments for pigeons." The boys also spent time playing arcade games at a sports club near the university. "I just lived over

Micro Users as Makers

the hill from there . . . and they had *Moon Patrol*, *Elevator Action*, *Defender* and *Donkey Kong* and that was the 'bee's knees' at that time," White recalls, adding that when he was about six or seven years old,

> Me and my brother, who loves games as well, we re-worked clones of arcade games. [Often] we made board game versions of them. The first game that we ever made together [on a computer], which was possibly the first game that Joseph made too, was a clone of *Moon Patrol*, with ASCII and written with no variables. . . . The way that we wrote the game was we started on it at first instance and then it's like you're writing a pick-a-path game, except that it's for real time. So we wrote the code just to draw a tank with the ground—just in ASCII, using text as the graphics. And then at the next frame, the world split into two possible events: either you stayed on the ground and a rock came closer or you jumped—jump was the only key. And then from those two events, we made two other events—if you had jumped again or if you hadn't jumped. So we were encoding every instance that there could possibly be in the whole world, rather than using variables.

Stories like these might seem remarkable now—frequenting a computer store every day after school for weeks, as Stephen did, not to mention an underage Symons playing games regularly in a pub for pocket money. The semipublic space of the computer store and the pub was owned by these boys: they made it their own. Along with the news agency, the sports club, and the university, these neighborhood haunts offered them either access to a computer or access to information about computing, and the chance to learn to program.

Gendered Access

While several scholars have researched gender in early workplace computing contexts (Abbate 2012; Hicks 2017) and demonstrated that the premicrocomputer-era teletype and arcade culture were often masculine (Rankin 2018; Kocurek 2015), we know little about the challenges women and girls faced regarding microcomputers at home. Although my sample size is small, there are still some obvious gender differences between girls' and boys' experiences of early microcomputing, particularly around access. Of my three female informants, two were girls when they learned to code. Both Fiona Beals's and Katharine Neil's coding endeavors were centered on the home, either theirs or their friends'. Though I have previously established—using photographs from the period—that girls were present in New Zealand arcades of the 1980s as players and participants, not simply onlookers (Swalwell and Bayly 2010), it is not surprising to hear that girls might have enjoyed less latitude to hang around some of the neighborhood haunts—where, as boys, my informants were able to gain access to the digital realm—than their male counterparts. For instance, when I asked Neil whether she grew up playing games in arcades, she responded "maybe once or twice but not really. [I didn't frequent] those dark and dingy places where people go to play games and wag school." She recalls, "My parents bought a BBC Master, which was after the BBC B, which was quite popular. No one's heard of the BBC Master, and no one bought one. My friends had Commodore 64s and Amstrads. . . . It took floppy disks that were actually floppy. It was just after the tape days. . . . Me and my mother learnt BASIC, the programming language."

In Neil's Auckland peer group, having a computer at home was seen as a status symbol. "When I was a kid, a computer was a luxury item for having fun . . . it wasn't something that you had in your house because you needed a computer," she explains. She attributes this as one reason why it was socially acceptable for girls to be interested in computers: "It was ok for girls to be interested in computers because it was a luxury item. If you could print out your school project on your parent's dot matrix printer, for example, it was like you were rich or something. And if it was one of those early laser printers, then you'd really 'made it.' . . . It wasn't nerdy at all. I don't think it had any stigma."

Neil recalls hanging out at her friend Naomi's house when she was about fourteen and doing things on the computer there, explaining that "they had a much better computer." She also reflects on girls' access to computers: "Definitely more boys seem to have got hold of computers than girls. I remember a boyfriend—this is when I was 20—telling me that his twin sister and him for their birthday, he got a computer and she got a dog. . . . Girls had to fight for attention in the classroom. And they had to fight to get on the computer at home."

Fiona Beals grew up in Westport, an isolated region of New Zealand, with little contact with others who were programming. Her network was thus very local. Like Neil, Beals remembers girls having differential access to computers:

> My mother had a boyfriend and he bought his kids a Commodore 64. That was where I sort of realised that there was an inequality about computers at that time. Only the boy used the computer, and he would never ever let any of the girls on. He would have friends around to play and they'd be on the computer. His sister

Micro Users as Makers

was there and she'd be allowed to play one or two games a day. . . . I think I got to play *Ghost Busters* once, but because I hadn't really learnt the strategy of the game, that once was a very short once. As a consequence, a couple of years ago I bought *Ghost Busters* (the Sega edition) on TradeMe, and now I know if you gave me it, I could play it. But yeah, it was really unfair.

Beals was another autodidact, teaching herself to program on a ZX Spectrum she'd received from an elderly couple her mother knew. She describes one of the programs she wrote this way: "The one program I do remember, which was totally fresh, was that I generated a pyramid on the screen that was firing bullets, simulated style, probably numbers and letters (I used to use x's to generate the picture). So I had this pyramid of x's, bullets that were x's, and aeroplanes that were flying across, and it was all random. When the bullet hit the aeroplane, the aeroplane would explode. But it was done in a randomised way, so that you never knew where it was going to hit."

Beals did not own a tape deck, so she had no way to save her program. "I remember that I left it on because it was such a great thing, and my computer overheated," she recalled. Because she couldn't move the computer without losing her creation, she showed it to the people who were around before it overheated, recounting that "they weren't really that impressed. No one else knew what you had to do to get there. Once you set it to run, people just thought that that was what you'd done. No one actually realised that I'd put some thought into this and I'd written this."

Beals's experience shows that even when girls managed to secure access to a computer, their achievements might not have been recognized or encouraged. The program has no name and only exists for a short while. It is—as Giard says of the language of home cooking—"modest": unlike with a restaurant meal, the cook "does not claim status as inventor or creator" (de Certeau, Giard, and Mayol 1998, 221).

Neil's and Beals's accounts are consistent with studies that found there was a drop-off in women's participation in computer science degree programs at universities in the 1980s. The percentage of women majoring in computer science peaked around 1984 and then went into decline. One explanation offered for this is that families were much more likely to buy computers for boys than for girls, even when girls were interested in computing. This translated into the level of knowledge that new computer science students were expected to have. As a National Public Radio story reported, "In the '70s . . . professors in intro classes assumed their students came in with no experience. But by the '80s, that had changed" (Henn 2014). There are no indications that 1980s Australia or New Zealand were significantly different from the US in this respect.

Dorothy Millard was at home looking after her children when she decided to buy a Commodore 64, simply because "I like new technology. I liked the idea, the kids liked the idea." She explains:

> Well, the Vic 20 was quite popular but these Commodore 64s were just coming out so I decided I'd go for this Commodore 64 and get the latest and greatest as it was at that time. And, of course, when I brought it home—I had small children—the two boys in particular were all over the computer and wanted to play with it. So they played with it during the day and then when they went to bed I learnt it. It was mine. While they were at school I would study it and learn how to program it. I never actually had any lessons.

While Millard's computing activities were located in the home, she was quite well networked. She ran a helpline, sharing her solutions to adventure games with people who were stuck. People would phone, having heard about her via word of mouth. Millard explains she had "maps of all the mazes and different things in the [games], and I would get my map out and my hint sheet and I used to tell people, help them with the bit they wanted help with. I enjoyed it very much. I was at home at that time with small children so it suited me well. It was something I could do at home plus it was a big interest of mine, of course." An example of one of Millard's maps and solutions dating from 1989 is seen in figure 3.3. It is a two-part guide to Darryl Reynolds's game *The Search for King Solomon's Mines* (1986), published under the Softgold label.

Millard had connections well beyond the neighborhood, indeed into the British adventure gaming scene through her involvement in the UK's Adventure Probe group, whose zines she received by mail. She also spent a year living in the UK, where she noticed differences between British and Australian Commodore cultures: "There were more magazines in England. The magazines at that time had the adventure help lines and all that sort of thing and I was involved with that. . . . I was involved in a group called Adventure Probe in England and I marketed my games through that in England."

We will hear more about Millard's activities later, but it's interesting to note that—echoing Neil and Beals—she also faced access issues. However, as an adult, she was able to resolve these by purchasing a second computer. She explains, "As I said, the kids loved [the Commodore 64]—they were all

Micro Users as Makers

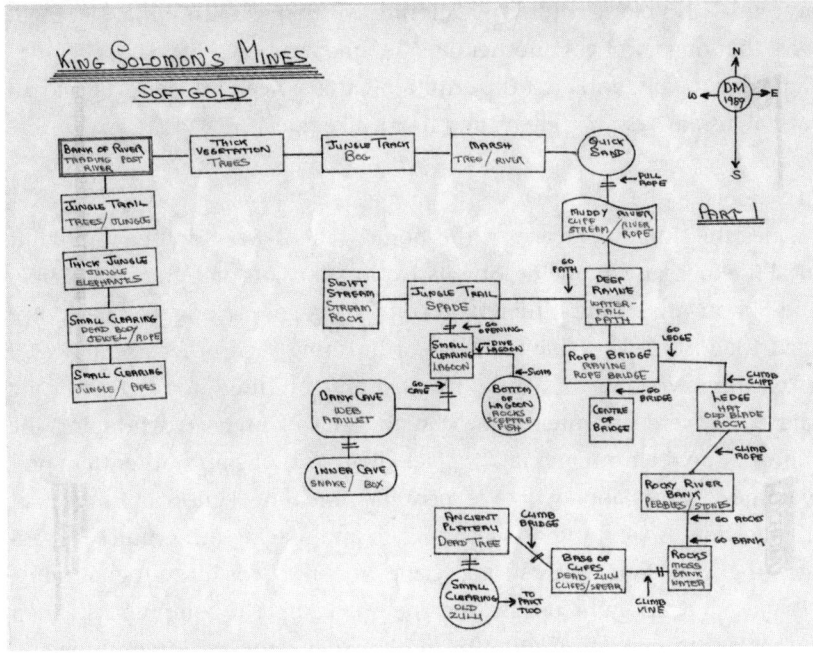

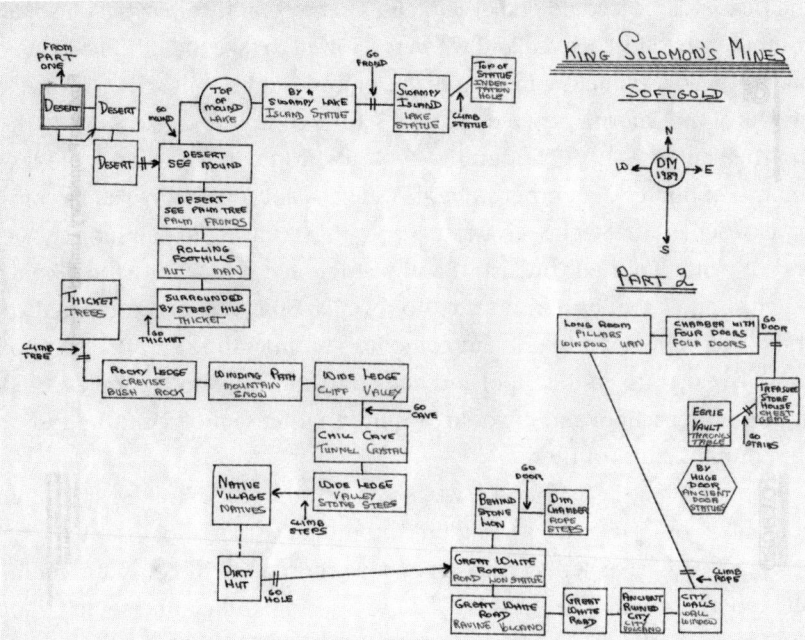

Figure 3.3
Dorothy Millard's solution to Darryl Reynolds's *The Search for King Solomon's Mines*. Courtesy of Dorothy Millard. (See plate 4.)

over it. It wasn't long before I got another one that was just mine. I let them have that one and I got another one. As time went on a little bit, the price came down a bit, and . . . I thought 'Blow this, I need my own,' and I had several disk drives for copying and things like that."

At School
Besides the domestic space of the home, school was another important local site for many of my informants' explorations of computers. Mark Sibly and Simon Armstrong's interest in computing goes a long way back, like their friendship. They have a habit of finishing each other's sentences. I asked them whether they were involved much in user groups. Armstrong said these were too much "like church," full of users "worshipping the thing . . . up the front and talk[ing] about all sorts of crap" rather than programmers. Sibly elaborated: "We spent far more time, I guess, at home or at friends' houses, learning how to program and making little games, experimenting with them." Armstrong relates how they got into programming: "It started at Selwyn College, and they had Apple II computers in their computer room, way back in 1985 or '86. They let us take them home on the weekends. You could book them out. That was pretty amazing: personal computers had just arrived and we were allowed to take them home. So my big brother—even before I got to high school—was bringing them home."

Few of my informants credit formal school computer classes with igniting their interest in programming. Typically, computer teachers were less important figures than other students were. As Sibly put it, "We had a computer teacher, Mr. Steel, who was never enthusiastic about the subject. We sort of worked out our own stuff and just ignored him." The school computer room was an important site, however, as Sibly recalls: "You could go there after school, and book out computers at lunchtime and that kind of thing. We used to go to school and hang out and, you know, people would go there after school and it would be quite an intense programming sort of time. You learnt a lot there."

Around 1988, Beals found a way to access her school's 286 generation computers by becoming a computer monitor (a student assistant to the teachers). This gave her license to spend lunchtimes in the computer room. She reports finding the early PCs "more limiting and cumbersome" than the Spectrum she had programmed earlier, and suspects that—apart from the computer itself—the institutional context had something to do with it.

The Spectrum she had messed around with at home had been hers, whereas the computers at school were everyone's.

School was not only a site for accessing computers and mastering coding techniques. For Nickolas Marentes, lessons—particularly in chemistry—provided precious time that he would divert to programming.[13] As he said to me, "I didn't need a computer to write the code. I just needed a piece of paper, and I wrote the code on paper and when I got home I typed it in." Marentes was coding in machine code by this time. A sample of his handwritten source code is shown in figure 4.8.

Up to this point, I've analyzed the way that micro users—youngsters and adults alike—operated on a predominantly local basis. Of course, the software and games they wrote did not always remain local, and distribution will be considered in the next chapter. For now, I want to turn to the second key theme I borrow from volume 2 of *The Practice of Everyday Life*, Giard's conceptualization of "doing cooking."

The Cooking Analogy

In part II, Giard hones in on domestic space, specifically the kitchen, which she calls "a theatre of operations for the 'practical arts,' for the most necessary among them, the 'nourishing art'" (de Certeau, Giard, and Mayol 1998, 148). Traditionally considered "women's work," cooking in the home has not been accorded much respect. Yet cooking demands "a subtle intelligence full of nuances and strokes of genius, a light and lively intelligence that can be perceived without exhibiting itself . . . *a very ordinary intelligence*" (158). Giard writes that cooking practices are "activities that people consider very simple or even a little stupid," but that when one considers what goes into it, cooking is actually quite a complex activity. Cooking demands that one calculate, evaluate, and then quickly realize. At one point, Giard even writes that cooking "calls for a programming mind" (157).

Cooking is an apt analogy for programming and homebrew software development more generally. Like culinary practices in the home, homebrew practices have typically not been accorded much respect. Like cooking, programming a game requires a "lively intelligence," and might also be considered a form of "work, without schedule or salary . . . work without added value or productivity" (159). I am not alone in seeing utility in an analogy between programming and cooking. Nathan Ensmenger quotes a

1967 *Cosmopolitan* article in which Grace Hopper compares programming to following a recipe: "Programming was 'just like planning a dinner,' the article quoted software pioneer Admiral Grace Hopper as saying, "Women are 'naturals' at computer programming" (Ensmenger 2012, 73). And, like those cooks Ferrier interviewed who minimized their meal creation as nothing special, some of my informants considered their activities unremarkable. Sometimes I got the feeling this was because they didn't think their efforts were worthy, but at other times it was out of a sense that "everyone was doing it." The sense of ordinariness or everydayness is, of course, ironic given how little attention homebrew has received to date. Yet there's also a paradox here, because some were also aware that not everyone was doing it; a number of informants either censored themselves or limited who they told about their practice to avoid the stigma they were aware could be attached to the activity of programming. John Passfield's first game, *Chilly Willy* (1984), was published for the Microbee when he was in tenth grade. He recalls that "some of my nerdy friends, they knew I was doing this because they had computers as well." Passfield wrote another game title, *Halloween Harry* (1985), in eleventh grade. He had changed to a different school by then, with, he explains, "different groups of friends there and I never told them what I did. . . . I don't know why. I guess I was just embarrassed about it because it was . . . very nerdy to make a game."

Giard recognizes that knowledge about cooking is transmitted in many ways: "Someone showed me how to do it; I saw it done this way; I ended up figuring out how to do it" (de Certeau, Giard, and Mayol 1998, 202). Sometimes one just ends up "figuring out how to do [something]" through experimentation. In chapter 2, I discussed the availability of books and magazines designed for use at the computer to teach oneself to program, and the prevalence of a discourse of hands-on, experiential learning. The ways in which such titles were used at the computer were not only experiential but also often highly *experimental*, involving much trial and error. For instance, Beals recounts that in learning to code she started by typing in other peoples' game programs. Then, she explains, "Once I clicked onto what was happening with the book, I was able to go 'well actually I don't want it to do this, I want it to do that.' I could start manipulating the code to do other stuff."

Much like Giard's observation about cooking, Beals's process was one of learning by doing, as she describes: "When I was writing the code myself, I would always write 5 or 6 lines of code and then end it, and test it, run it and see how it would go." "Doing programming" involved establishing by trial, echoing Raymond Williams's insight that experience was "once the present participle not of 'feeling' but of 'trying' or 'testing' something"; that is, of experimentation (Williams 1983, 128; Swalwell 2008a).

The use of how-to books and other resources provided a base set of instructions with which one could fiddle around and improve, improvising and trying out new techniques. John Perry describes how he came to understand BASIC as if he were "learning a new language." While his family had some books around that he would occasionally refer to, more often, he recalls, "you learn a 'word'—a function, a command—and you use that, and suddenly it changes all your programs, because suddenly you've got something that you can do. And then you learn some other trick. And generally I'd see someone use it in a program or something, and I'd look it up in a book sometimes, but mostly there was [no need for a book]. Mostly to start with you'd just copy the program and change a few little things to work out what was going on."

The linguistic metaphor Perry uses here is apt, but one could just as easily compare the sample text to a recipe. While recipes exist, often an ingredient or tool is not at hand, so one must *improvise*. A confident cook is not beholden to exactly following the instructions but will happily substitute. "Creative ingenuity" is required, Giard writes, as are "alternative ministrategies when one ingredient or the appropriate utensil is lacking" (de Certeau, Giard, and Mayol 1998, 157). Giard continues, in "doing-cooking . . . [often] the recipe itself loses significance, becoming little more than an occasion for a free invention by analogy or association of ideas, through a subtle game of substitutions, abandonment, additions, and borrowings" (201).

While cooking practices are passed on, sometimes from generation to generation, Giard observes that people also have their own ways of doing things. There's no one way to do cooking, so "each operator can create her own style according to how she accents a certain element of practice, how she applies herself to one or another, how she creates her personal way of navigating through accepted, allowed, and ready-made techniques" (156).

So it is with programming. It is possible to arrive at a number of different solutions to a challenge. Darryl Reynolds cites this variability as one of the appeals programming holds for him: "You could program it and what's more you could make it do the same thing umpteen different ways. And I think that was the thing that really got me about computers, that [programming] was open ended."

Homemade

The homebrewing of software resonates, of course, not just with the cooking of meals but also with the brewing of beer and other beverages in the home. Indeed, the *OED* gives the following definitions for the term:

> Homebrew, *n.*, 1. Home-brewed beer, wine, or other alcoholic drink.
> Home-brew, *v.*, To brew (beer, wine, or other alcoholic drink) at home.
> . . .
> B2. *Adj.*, 2. colloq. In extended use: home-made; improvised, amateur; esp. (of electronic equipment, computer software, etc.) designed by a user as an alternative or adjunct to a proprietary product.

That the *OED* recognizes electronics and computer software as domains of homebrew practice—with an example of usage from 1977 referencing ham radio antecedents—is entirely appropriate. Practitioners who brew their own beverages at home eschew industrial production techniques, preferring handmade, artisanal methods. The homemade aesthetic is appreciated. Variations—and the inclusion of unusual or unique flavors (beer with kaffir lime leaves, anyone?)—are celebrated. No two batches are alike.

The fact that many homebrew game developers were young has made it easy to dismiss their activities as marginal, lacking legitimacy. The diminutive term "bedroom coder" is one manifestation of this (Campbell-Kelly 2003; Caufield and Caulfield 2014). Though the designation has passed largely without challenge, I prefer the terms "home coder" or "homebrew developer" not only because "bedroom coder" is condescending but also because it misses the positive resonance of homebrew.[14] It is also inaccurate, as coding activities took place in various domestic spaces (see figures 3.1 and 3.2).

Making something oneself, de Certeau and Giard write, involves "overturn[ing] the imposing power of the readymade and preorganized . . . to trace one's own path" (de Certeau and Giard 1998a, 254). Paralleling the homebrewing of beverages, we might inquire whether the homebrewing of software produced unique "flavors" via the methods employed. Certainly, what was produced was individual, singular. The titles at least started as one-offs, custom rolled rather than commercially produced en masse. I will have cause to return to the cooking analogy in chapter 5 when I discuss so-called electronic circuit cookbooks.

Like brewing and cooking, homebrew programming also entails gift economy elements. Just as one might prepare a meal or brew to enjoy with others, homebrew authors were happy to share know-how among fellow coders, including instances of virtuosic programming and clever tricks. Programmer Cameron McKechnie recounted his involvement in the Amiga game *Sorceror's Apprentice* (1990), which he worked on with a group of his friends. "I conceived/designed the platform algorithm which I then shared with Mark [Sibly] when he said it couldn't be done" (personal communication, October 31, 2012). The sense of delight in his being able to prove his friend wrong is still palpable over email thirty years later. Other times, they'd share their games with their friends. Sibly distributed his game *Dinky Kong* (1984) (figure 3.4) to his schoolfriends in Ziploc-style bags with color artwork he did himself, using Letraset, under the name Perspective Software. Later, he swapped the game with the owner of a computer store for an external floppy drive. The computer store owner released *Dinky Kong* commercially, under the name Kiwi Computer. When I asked Sibly whether he was proud at cutting such a deal, Armstrong commented that there were no benchmarks or criteria: "It was all done in a bit of a void, so you didn't really know. You know, there was no one telling you that this was good or bad or anything like that."

Of course, "tracing one's own path" and being able to perform clever programming tricks were not universally appreciated. Nickolas Marentes did a TAFE (technical education) course in data processing after finishing high school. He remembers:

> They taught programming in BASIC, surprisingly on TRS-80 computers, which was a piece of cake for me. The thing that annoyed me the most was that . . . I was the one that the students were asking, "How do I do this? Can you fix that?" When the exam time came I got every question right on the exam test but the lecturer must have been from the last century because he didn't understand all of the programming tricks that could be done on those TRS-80 computers, and of course I was writing code that was utilising some really fancy sort of stuff, and they were marking me wrong for it. . . . We never had an opportunity to compare or to show . . . but they would tell us what we should have done, and that's where I knew, "Oh, hang on, I could have done that even better" [laughs].

Symons also found his talents somewhat underappreciated. Though still a school student, he had been publishing programming books for some years, and some of his books were actually in the library at his high school, which was the first public school in the state to get networked BBC micros. He recalls that his school was "held up as the pinnacle of what leading edge tech and education would be, but no-one there knew how to use them."

Figure 3.4
Cover art for Mark Sibly's homebrew game *Dinky Kong* (1983), as self-distributed. Courtesy of Mark Sibly. (See plate 5.)

Micro Users as Makers

Two mathematics teachers were picked to teach computer science and taught themselves to code from books, including Symons's books from the school library. Symons recalls that he almost did private tutorials with these teachers, both of whom left and got higher-paying jobs in the computer industry: "It was pretty funny. And one of them gave me—I'll never forget to this day—gave me a B in Computer Science as he left."

Satisfaction

Despite a lack of benchmarks or recognition, and sometimes little in the way of encouragement, my informants clearly derived a sense of personal satisfaction from their coding efforts. There is, as Giard writes, "a profound pleasure in . . . practicing a modest inventiveness" (de Certeau, Giard, and Mayol 1998, 213). Like "the preparation of a meal [which] furnishes that rare joy of producing something oneself" (158), informants report finding pleasure in creating games. This came out strongly in the interview with John Passfield. He enjoyed creating and predominantly thought about his game making as a form of creativity, stating that "plugging away over the Christmas break, you know, during the day and at nights [was] fun. I guess it was creative like writing a story or building something over the holidays; that's what it was about."

Having a game published was a high point for many. Even if they didn't feel able to share their achievement with others—as in Passfield's case—he recalls that "it was very exciting [and] I got the tape sent to me so I'd think it was real."

Symons found "the idea that I could be published" and "that there was more for me to learn" the driving force. The experience of working with publishers and editors was also very rewarding. He said, "I was honing a craft," adding that "there were a lot of skills I got for . . . free by having the editors and publishers who would send back their mods and tell me the basic rules in terms of English and structure and near-journalism; not journalism, but near, in terms of engagement, in terms of writing a book that's on what could be seen as boring subject matter, but trying to make it exciting, trying to make games more exciting than they are when they're text-based or, you know, dot meets dot, dot changes colour." Speaking about his book sales and royalties, Symons articulates the pleasure and reward of having produced something himself: "It wasn't lucrative. But it was something else. I'd be lying if I said it wasn't a fantastic boost to your ego to walk into Foyles or to walk into London shops and see your books. Because about

4 or 5 years later I was living in London and I'd walk down the street and see the books, and so there was a lot more than the monetary side of the thing. For me, I was surprised I was getting paid at all."

In chapter 4, I delve further into informants' motivations for their homebrew activities and detail how they thought about their practice.

Theorizing Homebrew

One reason why home coding may have been overlooked is its everydayness, its homeliness, its ordinariness. It wasn't glamorous or spectacular, and even when games were shown to those in the local neighborhood, many people didn't know what they were looking at. In the remainder of this chapter, I want to ask what is at stake in homebrew and what is at stake in restoring it to a chapter in the history of computing. What does it contribute to media and audience studies? What does a focus on the vernacular digital practices of homebrew authors make possible?

Following de Certeau and Giard in writing of ordinary culture in their 1983 report *L'ordinaire de la communication*, I propose that the three dimensions of user making they identify—the aesthetic, the polemical or political, and the ethical (de Certeau and Giard 1998a, 254–255)—help to articulate homebrew's significance. To address the political or polemical aspect first, that ordinary practices can have political significance has long been argued, in a range of disparate disciplinary contexts. That "the personal is political" has been an article of faith since second-wave feminism. De Certeau clearly envisages everyday practices as having political dimensions. Mark Poster writes that "de Certeau extracts consumption from theories of mass society and repositions it as a form of resistance" (Poster 1992, 103). Along with other foundational texts from the Birmingham and Frankfurt schools, de Certeau's scholarship has helped to shape cultural and media studies' attention to the politics of cultural practice. Regarding de Certeau, Poster continues, "His theory of consumption is an alternative both to the liberalism that bemoans the irrationality of mass culture and the Marxism that finds it always already controlled by the system. It provides a starting point for a type of cultural studies that is not predisposed to dismiss the billions of everyday practices in the late twentieth century" (103).

By and large, the political dimensions of cultural practice have been understood in oppositional terms. As Brian Longhurst notes, de Certeau's

The Practice of Everyday Life, introduced to Anglophone readers via John Fiske (and, I would add, Henry Jenkins), has typically been read as part of the incorporation/resistance paradigm (Longhurst 2007, 9). To wit, in reflecting on his own early writing, Henry Jenkins expresses ambivalence toward the metaphor of poaching he borrowed from de Certeau, noting that oppositionality came to occupy many academics writing on fan fiction: "Like all metaphors, 'poaching' enabled us to see certain things about fandom, offering a powerful counterimage to prevailing stereotypes of fans as passive consumers and cultural dupes; yet it also masked or distorted some significant aspects of the phenomenon, focusing on the frustration more than the fascination, *encouraging academics to read fan fiction primarily in political terms*, and constructing a world in which producer and consumers remain locked in permanent opposition" (Jenkins 2006, 37, my emphasis). De Certeau and Giard's flagging of polemical or political significance seems to me more nuanced than the way in which the concept of poaching from volume 1 of *The Practice of Everyday Life* came to be deployed; for instance, poaching always seemed to be about poaching commercial culture.

Roger Silverstone goes further. He notes that while "de Certeau's theories, and especially his consideration of the potential for resistance within the tactical times of everyday life . . . [provide] the intellectual space for a more strident populism," Fiske's reading of de Certeau is "an oversimplification" (Silverstone 1994, 162–163). Silverstone argues:

> There are at least two ways of reading de Certeau. One can find in his theories an opportunity to explore, and in exploring celebrate, the private, oral, poetic acts: the minutiae, the stubborn creativities, the potential transformations of public cultures, which mark and sustain our identities and our places in an overweening, increasingly imposing, contemporary society. Or we can recognise the scale and extent of that imposition, and see in the same activities a kind of superficial scratching, the equivalent to doodles on the backs of school exercise books, making marks but not affecting the structures, and intermittently (when our doodles are discovered) being punished for a lack of respect for the projects and structures, and above all for the authority, of legitimate institutions and values. (121)

The emphasis on a more ambivalent reading, on the minutiae and the private, might not be as dramatic as claims that consumption is political resistance, but departing from the fan, cultural, and television studies narrative that has focused on the politics of audience appropriations of popular media texts enables a different set of questions to emerge, about users' practices and making via the computer. Focusing on these smaller, not so grand narratives does not necessarily entail an acceptance of the status quo and does not neglect power; rather, the focus is at a micro level. Reflecting on de Certeau's remark toward the end of volume 1 of *The Practice of Everyday Life* that "it is always good to remind ourselves that we mustn't take people for fools," Giard offers that in de Certeau's "trust [in] the intelligence and inventiveness of powerless people . . . stands out a political conception of action and of unequal relations between a government and its subjects" (Giard 1998a, xxii). As de Certeau and Giard write about polemical or political dimensions of everyday practice, "the everyday practice is relative to the power relations that structure the social field as well as the field of knowledge. To appropriate information for oneself, to put it in a series, and to bend its montage to one's own taste is to take power over a certain knowledge and thereby overturn the imposing power of the readymade and preorganized . . . to trace one's own path through the resisting social system" (de Certeau and Giard 1998a, 254).

That de Certeau and Giard flag the aesthetic and the ethical dimensions of everyday practice—alongside the polemical—is important to note. Though the relative importance of each term—polemical, aesthetic, and ethical—is not explicitly spelled out, I read the three terms as interimplicated. Of the aesthetic, they write that "an everyday practice opens up a unique space within an imposed order, as does the poetic gesture that bends the use of common language to its own desire in a transforming reuse" (254).

Of the ethical dimension, they write that "everyday practice patiently and tenaciously restores a space for play, an interval of freedom, a resistance to what is imposed (from a model, a system, or an order). To be able to do something is to establish distance, to defend the autonomy of what comes from one's own personality" (255). In this articulation, the ethical and the aesthetic domains seem also to possess a political significance, the polemical an aesthetic importance ("bend[ing] its montage to one's own taste"), and so on. Each offers an alternative to what is "imposed"—whether this is a "power," an "imposed order," a "model," or a "system." Read together, each assists in articulating the significance of an ordinary cultural practice such as homebrew game development.

Homebrew authors learned how to use computers and used them to produce their own software. The results of their practice varied from knowledge and skill acquisition (Symons honing his writing craft and coding skills), to the joy of creating one's own game titles (Passfield's games or Symons's notebooks of code that he'd run in his head), to alternatives or adjuncts to proprietary products (Beals's ASCII pyramids firing ASCII bullets). Some, of course, wrote games with a political edge. I don't have any Australasian examples from the period, but others have provided examples, such as the 1980s Czech text adventure *The Adventures of Indiana Jones in Wenceslas Square in Prague on January 16, 1989* (Švelch 2013); "*Reagan*, in which the player had to dye the hair of the aging US president to stop him [from] starting a nuclear war"; and the satirical title *Denis through the Drinking Glass*, which commented on Denis Thatcher's drinking habits (Lean 2016, 186–189). But the stakes weren't always oppositional, nor the narratives necessarily emancipatory. Some found that the financial boost they gained from their hobby delivered outcomes with tangible personal significance in the way they were able to navigate their way through an "imposed order." Writing games in BASIC in one's spare time and selling them by mail through the local newspaper seems a humble endeavor for a microbiologist with a PhD running a hospital lab. Nevertheless, Arthur Streeter explained to me that the extra income he earned from his cottage industry activity "assisted us to pay off our mortgages early, and that was . . . a good thing; it's a weight to have removed. It gives you a chance to think about what you're going to do with your life too. I mean when you aren't a slave to the mortgage you can take risks and do things that might or might not work, which, when you've got a mortgage hanging around your neck, you know, you've got to meet those obligations." His selling of homebrew games during the 1980s enabled him, in the early 1990s, to give away his salaried position and take a "big risk," starting his own software company, which created autoanalyzer database applications for hospital environments.

The Disciplinary Stakes

If it has been considered at all, homebrew has been considered a private activity that lacks legitimacy with respect to a "productivist rationality" (de Certeau 1984, 69). The knowledge of amateur practitioners has often been devalued, and it has been assumed that practitioners/creators are not conscious of the knowledge they have: "the know-how of daily practices is supposed to be known only by the interpreter" (70–71). Like the television viewer de Certeau ponders in volume 1 of *The Practice of Everyday Life* ("it remains to be asked what the consumer *makes* of these images"), homebrew creators' activity is also quasi-invisible (31). Getting insights into ordinary computing culture requires asking, and listening to, users themselves.

Computer history is implicated here. While games have occasionally made it into histories of computing, ordinary people's uses of computers are almost always overlooked. Home coding—of all sorts—has apparently been deemed to be outside a history of "proper" software development; that is, of "serious" computing. Beyond a few exceptions, computer historians have shown little interest in ordinary cultures of computing, the popular, and the vernacular, apparently being singularly uninterested in reconstructing the lives of ordinary people. Why is this? Sometimes there is an elitism at work, where "insiders" marginalize the activities of amateurs. Their "literacy" might be a literacy of more sophisticated languages (cf. BASIC as common). Or they might (erroneously) believe that businesses and computer makers have a monopoly on innovation: "initiative [does not] take place only in technical laboratories" (de Certeau 1984, 167). I hope that this study of 1980s homebrew practices might contribute to a new vein of scholarship on the popular reception of microcomputers: of ordinary or unofficial culture; that is, "culture as it is practiced, not . . . what is most valued by official representation" (de Certeau and Giard 1998a, 251). Without wanting to romanticize the period, the homebrew case study concerns overlooked actors ("nerds," nonprofessionals), working at overlooked sites (the local neighborhood), on overlooked technology (microcomputers considered as toys), creating overlooked products (games dismissed as "clones"). The case study thus provides the opportunity to interrogate instances of vernacular consumption, remembering and recapturing something of the time when computing was new, and to grasp what it meant for everyday users. Studying homebrew reveals a history not of "great men" or great productions in computing but of schoolchildren and interested adults, with typically local aspirations, having a go, learning, and sometimes breaking through.

Moving forward, several larger questions remain in terms of the contributions of such a history. What part might "ordinary culture" play in a history of software? How might we begin to explicate the ways in which the legitimized and the delegitimized "enrich each other" (de Certeau and Giard 1997, 105)? And how might we account for the changes in legitimacy that occur over time? I take up some of these questions toward the end of the book, in a discussion of the contributions that homebrew developers continue to make to the remembrance of early microcomputing and digital heritage more generally.

What might a study of homebrew contribute to media, cultural, and audience studies? One of the themes of this book is the continuities—as well as the discontinuities—that homebrew gaming has with the longer tradition in cultural studies and cognate areas of theorizing audience activity. Ethnography and audience are concepts that are constantly present throughout this book, even if they are peripheral to my methods and chosen terminology. I write cognizant of the decades of ethnographic studies and theorizing by those who have examined how audiences consume a variety of media texts, from magazines, to films, to television, games, and beyond (Marshall 2004; Allen 1985; Gruber Garvey 2003; Rosen n.d.; Morley 2003; Taylor 2006; Banks 2013). I admit to some frustration when I read claims that position videogames as the medium of active participation par excellence, contrasted against other allegedly "passive" media that came earlier (e.g., Donovan 2010, xiii). Attending to the ambivalence and complexity of audience reception has been a constant feature of my scholarship (e.g., Swalwell 2008b, 2002). However, as I will show, there are moments where homebrew practices exceed the extant traditions of theorizing active audiences, points where homebrew practice genuinely cannot be contained within or adequately explained by received theories of consumption and reception, making clear the need to rethink these theories, particularly around technical user productivity. As I will argue in chapter 5, although we might speak of homebrew creators "writing" or "authoring" game and software titles, users' activities with microcomputers went well beyond content generation. Many were engineers in their own right, a fact that flags a potent discontinuity between homebrew and the ways in which histories of user activity have generally been conceived to date.

The homebrew case study also speaks to digital media histories. There has been a notable upswing in the scholarly interest being paid to the productivity of contemporary audiences, and particularly users of digital media. Everyone might be talking about consumers as active participants (Jenkins 2007, 361), but despite the warnings of some that "an emphasis on user-generated content as something newly technologically enabled . . . downplays 'a history of user-made websites, many of them fan-based, since the early days of the Internet' (citing Booth 2010) as well as a longer pre-Internet history of fan-generated material" (Hills 2013), there's been a strong tendency to focus on fan activities that cleave around the popularization of the internet and other new media from the mid-1990s onward. Before that, it seems to have been assumed that fans were working in an analog universe, literally cutting and pasting their zines together, for instance. Then along comes the internet, and a range of other forms become visible to fan researchers. The idea that digital fan productions might have predated the internet remains a strange absence within fan studies, an issue I will discuss in more depth in chapter 5. Homebrew game production in the 1980s disrupts any neat identification of the digital with the internet era, constituting an important corrective to accounts that (perhaps unintentionally) imply that digital media making is a recent phenomenon.

From another perspective, one of the questions that a focus on the vernacular digitality of the 1980s gives rise to is the extent to which it is possible to make and do today in ways that approximate the activities of 1980s homebrew developers. There are anxieties around the (lack of) opportunities to learn to code nowadays—what I call the "no one knows how to code anymore" thesis. For instance, in a *Salon* article, David Brin bemoaned the difficulty of finding a version of BASIC he could use with his son (Brin 2006). This is a complex and important series of arguments, and though I discuss a range of decline theses in chapter 4—notably the alleged decline of programming—a full explication of these debates is beyond the scope of the present volume.

Plate 1
John Perry on TVNZ's *Top Half*, with his game *City Lander*. "Computer Kid" segment, Top Half, June 1984, Television New Zealand, via Getty Images.

Plate 2
'Muzboz' at home in the family computer room. Photo by Brenda Lorden.

Plate 3
Andrew Stephen at his ZX81 in 1982, aged ten or eleven. Photo by Robin Stephen.

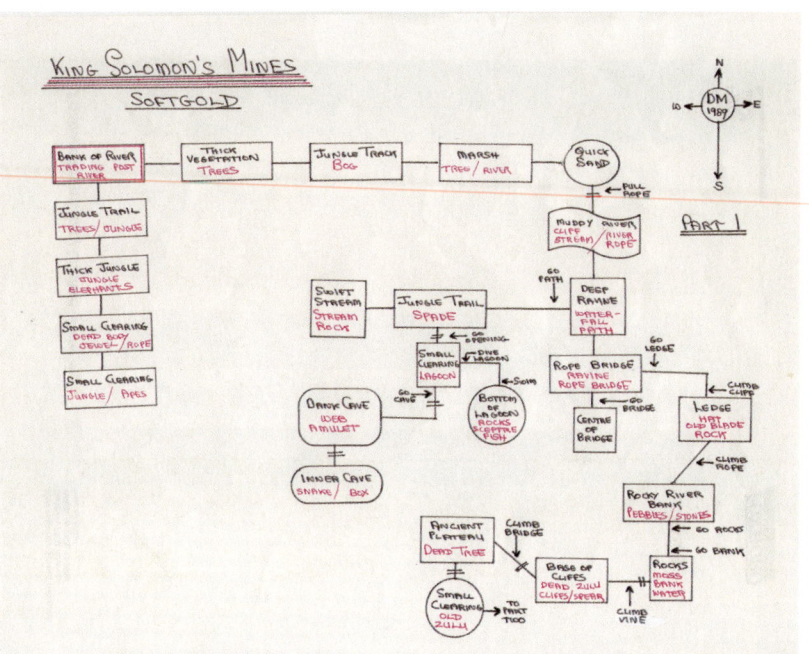
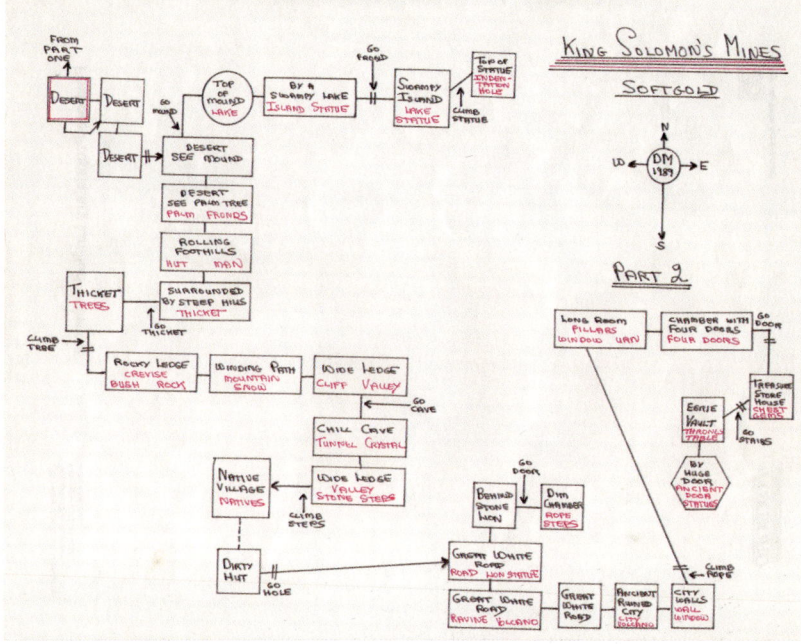

Plate 4
Dorothy Millard's solution to Darryl Reynolds' *King Solomon's Mines*. Courtesy Dorothy Millard.

Plate 5
Mark Sibly's homebrew game *Dinky Kong* (1983) cover art, as self-distributed. Courtesy Mark Sibly.

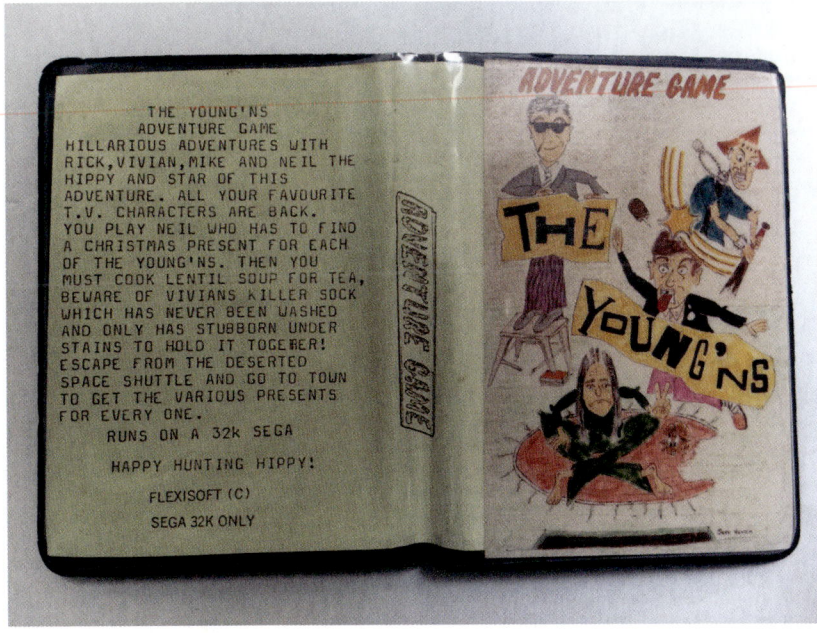

Plate 6

Tape cover artwork for Jeff Veitch's game *Young'ns*, published by Flexisoft for the Sega SC-3000. Photo courtesy of Clinton Rowe. The Retrowe Museum.

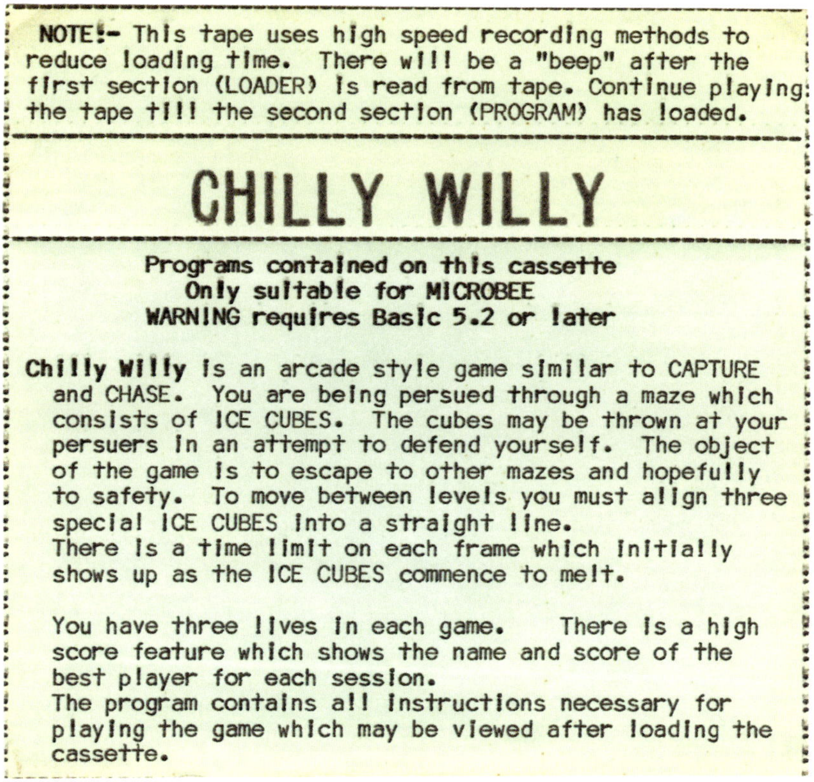

> NOTE:- This tape uses high speed recording methods to reduce loading time. There will be a "beep" after the first section (LOADER) is read from tape. Continue playing the tape till the second section (PROGRAM) has loaded.

CHILLY WILLY

Programs contained on this cassette
Only suitable for MICROBEE
WARNING requires Basic 5.2 or later

Chilly Willy is an arcade style game similar to CAPTURE and CHASE. You are being persued through a maze which consists of ICE CUBES. The cubes may be thrown at your persuers in an attempt to defend yourself. The object of the game is to escape to other mazes and hopefully to safety. To move between levels you must align three special ICE CUBES into a straight line.
There is a time limit on each frame which initially shows up as the ICE CUBES commence to melt.

You have three lives in each game. There is a high score feature which shows the name and score of the best player for each session.
The program contains all instructions necessary for playing the game which may be viewed after loading the cassette.

Plate 7
Cassette inlay with instructions for *Chilly Willy*. Courtesy John Passfield.

Plate 8
See following spread.

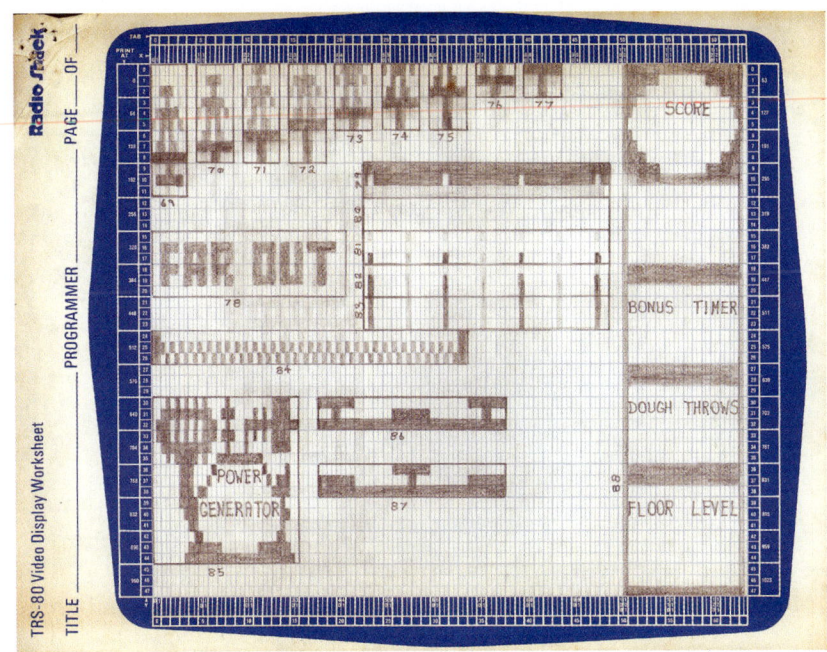

Plate 9
Graph paper graphics for a level in *Donut Dilemma*. Courtesy Nickolas Marentes.

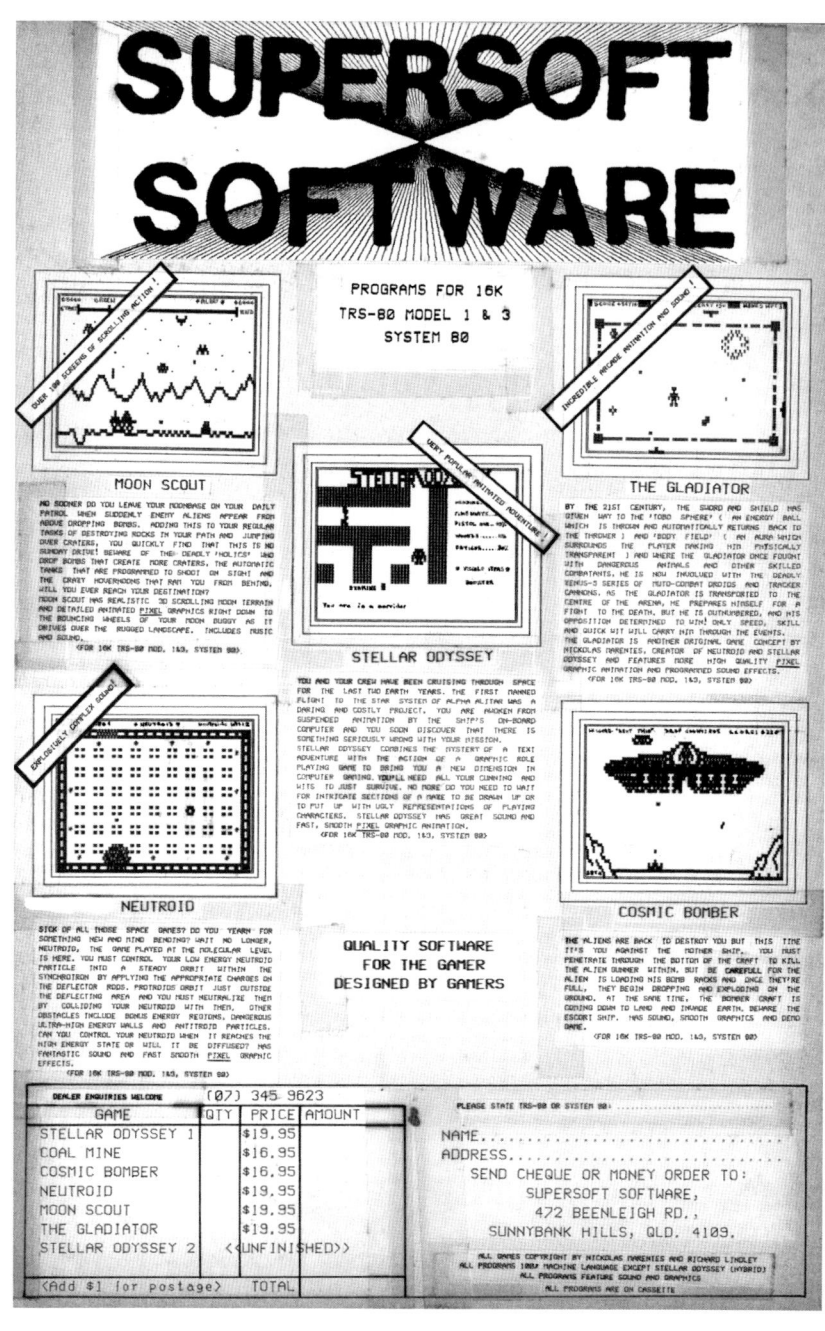

Plate 8
Supersoft Software sales brochure. Courtesy Nickolas Marentes.

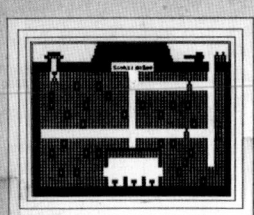

COAL MINE

THE GALACTIC STOCK MARKET IS EXPERIENCING A BOOM PERIOD AND THE PRICE OF PRECIOUS MINERALS HAS SOARED IMMENSELY THROUGHOUT THE GALAXY. YOU ARE A REBEL STAR PIRATE AND ONLY YOU KNOW THAT SOME OF THE LARGEST DIAMOND NUGGETS IN THE SYSTEM ARE HIDDEN DOWN THE BOTTOM OF AN OLD SHUT DOWN COAL MINE SITUATED ON STEITER 4. YOU MUST LAND YOUR CRAFT AND DIG YOUR WAY TO THE BOTTOM OF THE MINE TO COLLECT THE TREASURES. BUT HURRY, FOR THE MINE IS GUARDED BY PATROLLING DROME ROBOTS AND YOUR SHIP ABOVE IS UNDER ATTACK! WILL YOU GET BACK TO THE SHIP IN TIME OR WILL YOU END UP DIGGING YOUR OWN GRAVE?
(FOR 16K TRS-80 MODEL I ONLY AND SYSTEM 80)

STELLAR ODYSSEY PART 2

THE SAGA CONTINUES !

More software coming !

A WORD FROM SUPERSOFT SOFTWARE

The price of computer game software these days is expensive but what is worse is that their quality is no better. There are many game programs on the market selling for over twenty dollars that only use trivial graphic manipulation routines and even simpler sound synthesis routines. Our software is priced according to the quality of the program. Our latest games, 'Moon Scout' and 'The Gladiator' clearly show what we mean. More sound and graphic effects for your dollar. But good graphics and sound is only half the story. A unique and challenging game concept such as our program 'Neutroid' is just as important. And adventure games such as our 'Stellar Odyssey' (and now under development, 'Stellar Odyssey Part II') have about 50% of their development time devoted to story concept, 25% on graphics and sound development and the remainder to programming. We also welcome any feedback from our customers, whether it be a congratulations or a tip for future improvement on software design. In short, we believe our software to be excellent value for money and guaranteed to provide the gamer with many hours of enjoyment.
But we don't bother raving about our software,
We let our customers do it for us!.

MAIL ORDER BONUS OFFER

How would you like to recieve a free program of your own choice from our growing range of quality software with every mail order purchase?
Here's how it works:
Introduce a fellow gamer to a purchase of one of our programs then send us a combined cheque or money order (your order + fellow gamer's order). Also state which one of our other available programs you would like to recieve free and we will promptly mail them to you.

TRS-80 IS A REGISTERED TRADEMARK OF THE TANDY CORP.
SYSTEM 80 IS A PRODUCT OF DICK SMITH ELECTRONICS.

Plate 10
Homemade keyboard, collection of the Berlin Computerspiele Museum. Photo by Melanie Swalwell.

Plate 11
Homemade keyboard, collection of the Berlin Computerspiele Museum. Photo by Melanie Swalwell.

Plate 12
Select-a-Game, homemade console. Computer Archaeology Lab collection. Gift of Philip Kocent. Photo by Denise de Vries.

Plate 13
Cover of the *Microbee Hacker's Handbook*.

Plate 14
Pop*Star Pilot screenshot. Courtesy Nickolas Marentes.

Plate 15
Gate Crasher screenshot. Courtesy Nickolas Marentes.

Plate 16

Summer Games II (Catalonia patch to 4am crack) country selection screen (2017) by a2_poet. Courtesy a2_poet.

4   The Games

In the late 1970s and early 1980s, at what is the beginning of the digital age for most people, people are teaching themselves to code and writing homebrew games on low-end microcomputers. These users "get" computing and programming in a way that the wider public doesn't. Many of them are still at school, working alone and with limited resources, yet writing games and other software forms a big part of their everyday lives.

In chapter 3, I outlined the major aspects of the theoretical framework I use in this study, drawn from specific texts written by de Certeau and his collaborators that focused on users, practices of making in everyday life, and ordinary culture. I showed how their insights on the local and on cooking as a form of everyday improvisation could be applied to the endeavors of home coders. I argued that their concepts of users "tracing their own path" and "overturning the power of the readymade" provide a framework for evaluating the political, aesthetic, and ethical significance of such practices. This chapter is explicitly about what my informants made. In discussing their games and homebrew practice, it further fleshes out this theoretical framework with examples in informants' own words of what motivated them, of their satisfaction in getting the computer to perform as they wanted, and of the pleasure in creating one's own version of a game. The chapter is intended to reframe the terms of the debate away from claims that homebrew games were lacking compared to some purported industry norm (low quality and "disappointing" to their purchasers; see Campbell-Kelly 2003, 277) and on to a more positive articulation of the practice: a focus on what homebrew game development was rather than what it was not. I begin by introducing informants whose stories appear in this chapter before delving into how they understood their homebrew practice and how they articulate their motivations. I then consider homebrew authors' own criteria for writing games and what their influences were, detailing some of the games they developed before discussing where authors found their inspiration. I undertake a detailed excavation of Nickolas Marentes's game *Donut Dilemma* to drill down into one developer's practice, drawing on the archives he has kept. After establishing the ground of homebrew practice, the second part of the chapter returns to several allegations, considering the charges that homebrew games were just "clones" and that writing games oneself was "just a stage" in the reception of micros that people subsequently moved on from. The last part of the chapter considers the limit cases of those who set up more elaborate distribution networks and then reflects on some of the changes that the mid-to-late 1980s brought to computing and how these affected homebrew practice through to the end of the decade, including a critical consideration of the decline theses of Frank Veraart and Graeme Kirkpatrick.

The Informants

My interview sample comprises a total of seventeen informants who were home coders (eight from New Zealand, eight from Australia, and one from the UK), who programmed for a range of different computer platforms. I have already mentioned New Zealanders Simon Armstrong, Fiona Beals, Katharine Neil, John Perry, Mark Sibly, Andrew Stephen, and John White, all of whom were at school in the 1980s. (Selwyn Arrow makes his appearance in chapter 5.) Of the Australians, so far we've heard about the young John Passfield and Ross Symons. In this chapter, they will be joined by fellow students Vaughan Clarkson, Matthew Hall, and Nickolas Marentes. Dorothy Millard, Darryll Reynolds, and Arthur Streeter were all adults. The sample includes more males than females, which is consistent with informants' reports as to participation by gender. All were self-taught programmers, but the way in which they pursued their game development activities varied greatly, from those who dabbled as hobbyists to those who made it their business to write and sell software, albeit on a cottage industry scale. Subsequently, a good proportion (thirteen of the seventeen) would have careers in aspects of computing and software development. Their coding activity might also be plotted on a continuum from those who used software development kits (such as *The Quill*), to those who worked in the language BASIC, to more advanced programmers, who wrote in assembly language or machine code.

Curiosity and Fun

For most of my informants, the main motivation for their homebrew practice was the fun they derived from coding. Several explicitly answer the "why" question by deploying a discourse of curiosity and fun. Ross Symons, for instance, reflected on the pleasures of producing, and of attaining mastery over the computer. Symons wrote books of code for Tim Hartnell of Interface Publications[1] (e.g., Symons 1984, 1985), instructing others how they could learn to program:

> In terms of writing books, I used to set myself ten pages a day, so I'd come home from school and try to bang ten pages out. I probably averaged eight or something like that, so you get a lot done, and I think the reason I was able to do it is because it was fun.
>
> Mastering these things is a lot of fun, you know, it's a battle of wits, you and the machine. They were good times. . . . I would hazard a guess that most people are like that that were in there at the time. You know, that's what would have driven them. I don't think I met anyone that ever said to me, "I did this for money" [or] "It was the financial reward that drove me."

Echoing Symons, Dorothy Millard repeatedly invokes her need for a challenge, and her hunger for knowledge, understanding, and mastery:

> I just used to buy every magazine I could lay my hands on so I could learn. I was just eager for knowledge. I just wanted to know how it worked.
>
> I rather like the text [adventures]. I like the puzzles in them. I like doing the puzzles so I did those for a while and I thought after a while that I'd like to have a go at writing one. I used *The Quill* initially and I wrote a few games with *The Quill* and then I also wrote some in BASIC just to show that I could, just for a challenge really. I always needed a challenge.

Millard then goes beyond the idea of writing games for the challenge of it. For her, it was also about working within the constraints and getting the computer to do what she wanted it to do. She continues,

> I was pretty good at hacking into the games. Even if they didn't want me to, I was good at getting in and accessing the code. Sometimes you'd get a game and it was so badly written you'd sit there and you're trying to work out what words it understood. The parsers were dreadful. Mine were never brilliant because memory restrictions were always a problem, but I used to try and get a good variety of words, as much as possible within the restraints of size. It was always a bit of a, you know, on the plus and minus side because of memory restrictions with the 64. . . . I really enjoyed doing it I must admit.

> I only ever wrote for fun really, I just enjoyed doing it. I liked making the computer do what I wanted it to do. I still do. When I use graphics now I like making the graphic appear: I want it right in the corner and that's where I'm going to put it. It's just a sense of making it do what I want and being able to program it.

Millard's explanation of her motives for writing these text adventures are reminiscent of de Certeau and Giard's conception of the multidimensionality of everyday practice, discussed in chapter 3, namely the polemical ("bend[ing information's] montage to one's own taste"), aesthetic ("bend[ing] the use of common language to its own desire"), and ethical ("tenaciously restor[ing] a space for play, an interval of freedom") facets.

Creativity

Others invoke creativity—as well as fun—to account for their game development endeavors. As John Passfield says, "I'm more of a content creator, a designer, than I am a programmer," and Passfield situates his early digital game design and development in a longer trajectory of creative practice:

> I've always liked games and making games. I've got—I didn't bring them in but I have—little board games which I made up. So the computer was an extension of that, and I loved storytelling too; a big thing for me was creative writing, as a kid. That's I think one of the reasons why I did the *Star Trek* game. It's a story, it's a narrative about escaping a space ship that's going to blow up and the little chances that, you know, you type things in the computer and it gives you a bit of story back. And that was probably again that idea of creating stuff [which] is why I went from *Chilly Willy* to *Halloween Harry*, to make something more original and put my own sort of spin on things.

Passfield showed his notebooks during the interview, and it was clear that people's coding labor didn't always result in finished products. But, as he showed them, his explanations of his characters and story ideas strongly conveyed the "profound pleasure" Giard says derives from a "modest inventiveness." He explained that he wrote the first iteration of *Halloween Harry* "plugging away over the Christmas break, you know, during the day and at nights. [It was] fun, so yeah . . . I guess it was creative like writing a story or building something over the holidays; that's what it was about." Like Millard, it is possible to read in Passfield's articulation of the pleasures he found in creativity an evocation of de Certeau and Giard's polemical, aesthetic, and ethical dimensions: "overturn[ing] the imposing power of the readymade . . . to trace one's own path" (polemical), "opening up a unique space within an imposed order" (aesthetic), and "defend[ing] the autonomy of what comes from one's own personality" (ethical) (de Certeau and Giard, 1998a, 254–255).

Just as games were not always finished, they were not always published, and Millard's oeuvre is a good example here. In the interview, she rattled off some fifteen game titles, only one of which, *The Dare* (1989), was published commercially. The rest she has as D64 files, which will run in an emulator. They are freely available online.

Passfield wrote prominently in one of his notebooks "BE ORIGINAL," and he nominates originality—being able to put one's own "spin on things"—as one of the rewards he found in creating games. Originality is one of the criteria some homebrew authors nominate as being important, and I will return to it in a moment. In the case of text adventures, that a game had a good story and was bug-free and able to be completed mattered. As Millard said of Reynolds's games, "They were great fun [and] very popular. . . . They were good because they were well written and they were complete-able." The other major criterion that mattered to informants was the quality of the programming, a key consideration particularly for fast-paced action games, as it affected the speed at which a game would run. For those whose primary interest was in the programming, if they felt they'd solved problems and done something well, then that delivered a sense of accomplishment as well as potentially appreciation by peers. While these qualities were valued by homebrew developers, they were not always attainable. Trade-offs were necessary.

Influences

As already intimated, homebrew authors often wrote games based on what they knew. Their influences were many and varied, but often came from the local milieu in which they found themselves: ordinary culture, local places, and various forms of popular culture—including other games—were grist for the mill. I am aware of several games that deploy Australian idiomatic expressions, "in jokes," and national icons that probably do not translate very well, reflecting their creators' sense that they were developing for a local—or at most a national—audience who would get the humor. Such titles include Streeter's game *Mozzie Zapper* (1987) (named for the ultraviolet light device that kills annoying insects such as mosquitoes in summer) and games featuring native fauna, such as Mytek's *Emu Joust* (1983) (jousting while riding large flightless birds) and *Bunyip Adventure* by Ross Williams/Grotnik Software (1984) (a bunyip is a mythological creature from the Australian bush).

Locality and everyday savoir faire frequently found their way into informants' games. Millard acknowledged that she drew on locations that she knew or had visited for her text adventures: "*Harboro* was based on a place where my mother lived at that time, Market Harborough in the UK, which is in the Midlands . . . but you couldn't say it's very realistic. There's a lot of imagination in there." Millard explained that the game *Canberra Canberra* was "base[d] on my visit there, yes, the places and everything. Plus I would get a map out and have a look." She liked the fact that there was "an Australian subject line" and noted that she had planned "to write more of those but just never did." Meanwhile, Marentes based his most well-known title, *Donut Dilemma* (1984), on his observation of what sometimes went wrong with the machine in the donut kiosk his family ran.

Popular culture was an important stimulus for many homebrew authors and supplemented the local influences with transnational elements. Films, for instance, provoked astonishment and excitement at futuristic graphics, as well as providing themes, characters, and story lines that inspired the games that homebrew authors developed. Vaughan Clarkson recalls that the computer graphics of the era made a big impression on him: "I guess I was influenced by some of the films that were coming out around that time. Particularly, I remember films like *Battlestar Galactica* . . . and even actually *2001: A Space Odyssey*, which had computer displays that I found amazing, and I wanted to be a part of that."

The cover art inside the cassette box in figure 4.1 attests that television shows—such as the very British cult comedy *The Young Ones*—were also source texts. The show was broadcast from 1982 to 1984 in the UK but like many television series enjoyed multiple reruns in other markets, including Australia and New Zealand. While the game clearly references the TV show, it is interesting that the creator, Jeff Veitch, opted to change the title slightly (*Young'ns*). Perhaps this was at the urging of the publisher, Flexisoft, who thought a slight title change would avoid intellectual property claims; they put a (C) after their name, indicating they were claiming copyright. In another early case of transmedia marketing, Reynolds remembers being asked by a distributor to quickly write a game that would tie in with the release of the film *WarGames* (1983) starring Matthew Broderick: "I went

Figure 4.1
Tape cover artwork for Jeff Veitch's game *Young'ns*, published by Flexisoft for the Sega SC3000. Courtesy of Clinton Rowe, The Retrowe Museum. (See plate 6.)

flat out on that and we produced it on the Vic-20 so that's how long ago it was. [Then] just before it was about to be released on the market, the people that were distributing the film said, 'No, we don't want you to use that name.' So we had a mad rush and changed it around and renamed it *Thermonuclear WarGames*. . . . That was done on the C64."

Fantasy and sci-fi novels, pick-a-path and "Choose Your Own Adventure" novels, comics, and space adventures are also variously nominated by informants as providing inspiration, demonstrating games' imbrication in a wider visual and literary culture in the era.[2]

An Ecology of Games

Apart from films, books, and television, games themselves—both analog and digital—were an important source of ideas and inspiration for homebrew authors. Early digital games existed in multiple forms—arcade, console, handheld, and microcomputer—and these comprised an ecosystem of sorts. While Erkki Huhtamo has suggested that prejudices against arcade gaming were one of the reasons for the "breakthrough of home gaming" (Huhtamo 2005, 15), this explanation is not very persuasive when it comes to microcomputer use or game development in Australia or New Zealand: it was not an either/or thing. Moral panics around arcades were a reality, but as I have previously established, in New Zealand, arcade games were located not solely in arcades but also in numerous other everyday sites, such as fish and chip shops, dairies (corner stores), swimming pools, and transport hubs (Swalwell and Bayly 2010). For instance, figure 4.2 shows at least eleven students watching one person play at the interislander ferry terminal in Wellington when they turned up two hours ahead of a ferry sailing. The siting of arcade machines—at least in New Zealand—made them an everyday phenomenon that people routinely came into contact with. Therefore, people were quite often motivated to write a game for their particular brand

Figure 4.2
Wellington interislander ferry terminal—"The space invaders machines came in for some heavy use at the Wellington ferry terminal building early yesterday morning when a group of Evans Bay Intermediate pupils turned up two hours early for a ferry sailing." Dominion Post Collection; Ref EP/1988/1725-F. Alexander Turnbull Library, Wellington, New Zealand.

of micro that was similar to a game they had seen either in arcade form or on a different brand of computer.

Clone Allegation

Games that look or play the same as other games are generally referred to as *clones*. The word has all sorts of connotations, none of them particularly positive. To be labeled a clone implies that a product is a rip-off in some way, a copy that is too close to the original to be distinguished in any meaningful way, a derivative product. To our contemporary sensibilities, it also raises legal questions of intellectual property and its infringement (although copyright in software was still being worked out at this time in Australian and New Zealand jurisdictions). Clones—or the authors of clones in this period—are often deemed to have failed for one of two reasons, either because the game bears a strong similarity to another game or because it's been thought that the gameplay was not good. When presented like this, the issue seems simple enough. Under closer examination, however, these grounds start to look shaky, the issues less clear-cut. As I will argue, it is worth moving beyond the dismissive reflex that is implicit in the term *clone* to consider the significance of making one's own version of a game at this time.

Arcade Influences: *Chilly Willy*

The influence of arcade titles on the homebrew author is evident in the first game John Passfield ever sold. *Chilly Willy* is a version of the arcade game *Pengo* that Passfield wrote for the Microbee, an Australian-made line of microcomputers. The tape inlay in figure 4.3 displays the instructions for play. The loading screen displays the rules of play, which are fundamentally the same as those of *Pengo*:

> Oh no! Chilly Willy has a problem.
> The evil Ice-Lord has sent his nasty henchmen to invade Chilly's home and destroy him, so he needs your help!!
> You must help Chilly to escape from the baddies:—
> Ice-Bot   Freeze-Droid   and Polar-Cat.
> By lining up the three Ice-Stars, Chilly can move onto the next pattern, and hopefully to safety.
> *** Hit any Key to Continue ***

> **NOTE!**- This tape uses high speed recording methods to reduce loading time. There will be a "beep" after the first section (LOADER) is read from tape. Continue playing the tape till the second section (PROGRAM) has loaded.

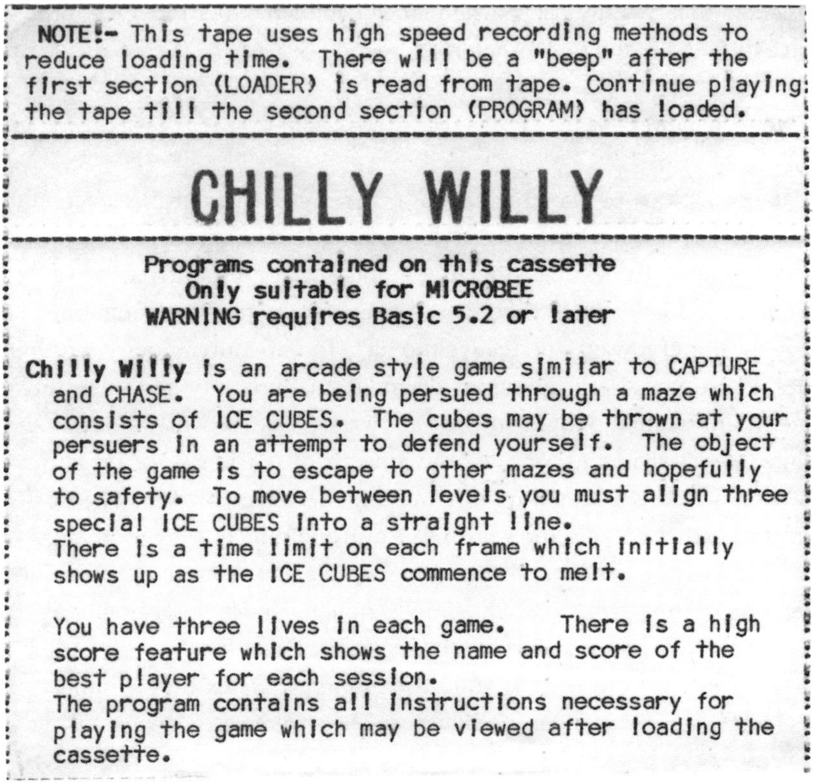

CHILLY WILLY

Programs contained on this cassette
Only suitable for MICROBEE
WARNING requires Basic 5.2 or later

Chilly Willy is an arcade style game similar to CAPTURE and CHASE. You are being persued through a maze which consists of ICE CUBES. The cubes may be thrown at your persuers in an attempt to defend yourself. The object of the game is to escape to other mazes and hopefully to safety. To move between levels you must align three special ICE CUBES into a straight line.
There is a time limit on each frame which initially shows up as the ICE CUBES commence to melt.

You have three lives in each game. There is a high score feature which shows the name and score of the best player for each session.
The program contains all instructions necessary for playing the game which may be viewed after loading the cassette.

Figure 4.3
Cassette inlay with instructions for *Chilly Willy*. Courtesy of John Passfield. (See plate 7.)

Passfield's level designs, enemy, and avatar models are different from those in *Pengo*, but the basic layout of the game and the objective remain the same. Passfield wrote *Chilly Willy* after playing *Pengo* in the arcade in the small New South Wales town of Kyogle. It was 20 cents to play it at the arcade and he loved playing it, so over the Christmas break he started writing the game, thinking that he'd replicate it so he could play it for free. As Passfield remembers:

> I didn't set out to make *Chilly Willy* to sell; I just set out to make it to play it at home because it was the challenge of trying to see if I could make what was in the arcades and I could play it for free. . . . It wasn't exactly the same as the arcade; I'd changed the levels around and stuff, not because of copyright, just because I was a bit more creative . . . and then, I don't know why, but at some point . . . I had

the address for Honeysoft at Gosford, and . . . I did a tape up and sent it down to Honeysoft [to consider for publication]. And I called it *Chilly Willy*, not realising that it was based on the cartoon character Chilly Willy, and I had no idea about copyright, being in high school. So I ripped off a game from Sega and copied the character's name from a cartoon and sent it down, and then they wrote back saying, "Oh, we like it. We'll publish it."

While versions such as Passfield's *Chilly Willy* are derivative, it needs to be remembered that variation is at the core of both the production and experience of 1980s digital games.[3] There are a multitude of examples to support this. Producers released arcade titles for home consoles from the earliest times.[4] And the reuse of game mechanics was common in the commercial sector. For example, the first part of the game *Horace Goes Skiing* (1982), created by Australian studio Beam Software, is—in terms of game mechanics—clearly *Frogger* (1981).[5] Early game producers were no respecters of originality, with ports[6] and clones central features of the games business. What, then, did the terms *original* and *copy* mean at this time?

Riffing on Benjamin

In media studies, the go-to reference for considering issues of originals and copies is, of course, Walter Benjamin's essay "The Work of Art in the Age of Mechanical Reproduction." In 1936, the German theorist wrote of the transformation the notions of "original" and "copy" were undergoing in the wake of technical changes. For Benjamin, the changes that mechanical means of reproduction such as photography and film had wrought on the work of art and the masses' desire to get hold of things up close by way of reproductions exacerbated the decay of what he called "aura." If we think about it, videogames have even less claim to being singular items with a "unique existence in time and space" than the objects on which Benjamin was reflecting (Benjamin 1992a, 214–217).

We can extend Benjamin's insight that "the work of art reproduced becomes the work of art designed for reproducibility" (218) to recognize that not only was the code of early videogames able to be copied but that at a time when versions of a popular arcade game would quickly appear for home consoles and programmable microcomputers, it was expected—or even required—that games be ported. It is now recognized in the game industry that innovating entirely new games is not only hard but also may not make commercial sense. In a blog post, "Originality in Mobile Games," Better

World studio writes that it can be risky to develop something 100% original, in that it can be difficult for people to learn the controls. Sometimes, they suggest, it is better to improvise and innovate on what already exists:

> All games are in a way or another a modified idea of some other game. Many of them even have the exact same mechanics, but only a few elements are different: it can be either the overall theme, story line, graphic style, mood and so on. Some games come up with new mechanics, but are still heavily based on everything else that's on the market already.
>
> What I'm trying to say here is that it's very difficult (and rare) to come up with a totally new genre altogether. Think about it like this: even game studios that have released a game that is very original and different from everything else still release other titles that are not that original.
>
> So the whole point is this: don't try too hard to create something so new, so unique, with mechanics that no one else has ever seen before, with controls that require [them] to learn so many skills that will amaze your players. That might even have a negative impact for your audience. As long as you come up with new ideas that improve and add to an existing genre, you're going to be fine. Think of new upgrades you could add, new add-on mechanics you could implement, or even combine elements from 2 games of different genres. (Better World Studio 2015)

We live in a post-Warhol era, in which new works often leverage existing ones. Musicians cover each other's songs, and aspiring artists sketch masterpieces. Yet there's a kind of pretense—or a willful denial—that such imitation doesn't happen every day. Developers are inclined to borrow and embroider familiar motifs, yet 1980s home coders are castigated for doing the same thing.

Writing a clone was not aberrant behavior but instead a widely accepted part of early games and computing culture. In terms of a user's experience, re-creating a game played at an arcade (or on another computer system) on one's BBC, Sega, or Microbee at home was actively encouraged; indeed, there were whole books dedicated to it, such as *Astounding Arcade Games for the John Sands Sega* (Love and Hancock 1984). Computers also often shipped with clones, as TestSheepNZ recalls regarding the ZX Spectrum: "The ZX Spectrum came with a manual and a sample tape which included a few demo programs. No-one really knew what was going to be popular, so included on there were a few educational programs (if I remember mathematically simulating fox and hare populations as well as Conway's life algorithm), together with a couple of games (a *Space Invaders* and *Pac-Man* clone)" (TestSheepNZ 2014). The incompatibility of 1980s microcomputers meant that many different versions of the "same" game existed.

The Games

Variation: *Hoards of the Deep Realm*

System incompatibility acted as a spur to homebrew authors programming clones, as Vaughan Clarkson's story clearly demonstrates. Clarkson was a keen schoolboy programmer and—like Passfield—two of his games, *Gridfire* (1983) and *Hoards of the Deep Realm* (1985), were published by Honeysoft for the Microbee. *Gridfire* was a version of *Crossfire* from the Apple and sold somewhere in the vicinity of 1,700 copies, while *Hoards of the Deep Realm* was a reworking of *Lode Runner*. As Clarkson put it, no one was going to port this to the Microbee computer, so you sort of felt "entitled" to do it. He recalls:

> I loved programming. I thought that I had a sort of natural skill there and I spent enormous amounts of my free time just trying to write things, trying to replicate things that I'd seen on other platforms, mostly the Apple II . . . which I thought was producing lots of good stuff. . . .
>
> [It] took me a really serious amount of time to write *Hoards of the Deep Realm*. I think that was probably a year and a half or two years of most nights and good parts of weekends cutting code and making it work. And you know I was really pleased with the result. I thought I'd employed quite a few tricks that I hadn't seen other people using to make a reasonably smooth animation. It really sort of, I felt, pushed the capabilities of the Microbee to its limit to make this thing work. But it had taken me too long to get there.

Hoards of the Deep Realm (1985) seemed to sell quite well, but it was published late in the Microbee era, by which stage the company was on the brink of collapse. The first royalty check Clarkson received for the title—which was in excess of $1,000—bounced.

Rather than terming his games clones, Clarkson—now a computer science academic—prefers to call them "reimplementations." He explains:

> I was interested in how games worked and I had taken games that I liked, figured out what all the game mechanics were and then thought "well can I actually do that—take the thing that works on the 6502-based Apple and make that run on the Z80-based Exidy Sorcerer?"[7] But there wasn't, apart from a reverse engineering and re-engineering effort, there wasn't too much—I'm sorry to say—of a creative effort that went into it from my point of view. I didn't work out new game dynamics and that sort of thing. Afterwards I thought I should have done that, that would have been a good thing to have done but at the time as a young high school student, I was just trying to figure out how these things worked.

These originally coded variations also made a sort of sense in terms of the economics of the early software industry, which—apart from big players such as Apple and Commodore—operated more on a national rather than an international footing. Passfield's and Clarkson's Microbee titles exemplify this, and sometimes, in the course of reimplementing a game, homebrew developers improved on it, working in some of their ideas. This was the case with *Hoards of the Deep Realm*, in which players could design their own levels.

While some cloned games were probably awful and sank without a trace, it doesn't follow that just because creators were amateurs, their games were of poor quality. Some homebrew titles were apparently very good. Their creators might have been young, but they approached their task with insight and discipline. The Play It Again research team has received comments on both Clarkson's and Marentes's titles via the Popular Memory Archive that indicate they were well received. Alan Laughton writes, "*Hoards of the Deep Realm* was simply one of the best Microbee games ever, in my opinion. One of the best things with this game was that you could design your own level(s) to make your own version of the game" (Laughton 2013). Meanwhile, "ross" said of *Donut Dilemma*, "This game was pretty amazing at the time. We used to spend ages load[ing] this in from our Tape Drives on the OLD TRS-80 and spent hours at a time playing it. One of my favourites for that time period" (ross 2014).

Both Passfield and Clarkson seemed a bit sheepish explaining to me what they'd done as teens writing clones or reimplementations. But another informant, John White, normalizes the practice, going so far as to suggest that, for home users, the writing of clones was expected: "That's like a typical thing to do. I think most kids when they start making games, when they're young, and looking at games that kids are making now as well, they don't actually [write something new] . . . making clones is actually the thing to do. People don't start by thinking up their own thing, they just like to make re-creations first, which is kind of nice, because any new game is ultimately a variation anyway. It's nice to just concentrate on what's around already."

Re-creating a game played at an arcade on one's BBC or Sega at home was the norm rather than the exception. Apart from enabling a programmer to hone their skills, writing a clone was sometimes the prelude to creators developing their own game concepts. This is demonstrably the case for Passfield,[8] who is a well-known figure in the contemporary Australian game industry, having developed numerous titles with original IP (Intellectual Property). During our interview, Passfield talked my colleague and I through his notebooks from the 1980s, which contained various designs, sketches, ideas, and scripts. As already mentioned, one of the pages of his notes contained the instruction—written to himself—to "Be original." Though he never received any criticism for *Chilly Willy* and he didn't realize what he'd done until years later, he felt that his teenage self had wanted to "make stuff that was different." He explains, "I did want to make stuff that was different . . . definitely something I wanted to do was create a new look or a new feel for something. . . . I wasn't content in just making, redoing another person's game." The experience of Mark Sibly—the author of *Dinky Kong*—whose Blitz Research would later go on to create various programming products (e.g., Blitz Basic, Blitz3D), resonates with White's comment, suggesting that "trac[ing] one's own path" (de Certeau and Giard 1998a, 254) in the game industry might entail following another's, at least in the early stages.

Given the frequency of reimplementations in the early micro era, the language and notion of originals and copies does not seem all that helpful. What I find more significant and interesting than the fact of widespread cloning is what seems to have been something of an imperative to make variants of games. John Perry's reimplementation of the Commodore 64 *Lunar Lander* game he'd seen provides further insight into the dynamics of borrowing and circulation of gameplay concepts. Perry wrote a version for the Sega SC3000, which he called *City Lander* (1984). But *Lunar Lander* has a much longer history. The Wikipedia entry, for instance, credits Atari as the developer of the 1979 arcade version, before calling it a "subgenre" and providing background on the 1969 version that launched the genre. As Benj Edwards tells the story, it was written by Jim Storer, but the comments on Edwards's article demonstrate the many systems the game was implemented for (Edwards 2009). David Ahl rewrote the game and included it in his famous book of BASIC computer games. Harry McCracken comments on Ahl's *BASIC Computer Games* that, "Like folk songs, its programs felt like part of a shared cultural heritage. They were passed around, mutating into multiple variants as they did so" (McCracken 2014). The idea that game themes and concepts comprise a shared cultural heritage on which one embroiders is reminiscent of oral traditions of storytelling. The concepts of orality and circulation are, I propose, far more productive ways to think about the issue of game variations than are notions of originals and copies.

Marentes and *Donut Dilemma*

While coding one's own version of a well-known game was virtually a rite of passage for the 1980s homebrew developer, the oeuvre of Nickolas Marentes is noteworthy, as he went in the opposite direction. Rather than proceeding from clone to original IP, he created several highly original games from the start of his homebrew days before finally writing what he called a *Pac-Man Tribute* in 1997 for the Tandy Color Computer 3. Indeed, Marentes attributes the poor reception of his third game, *Neutroid* (1983), to its being "too original."[9]

While he is not representative or in any way typical of the informants in my sample, Marentes has written an account about the development of each of his games. He has also kept exhaustive archives. Documentation related to each of his games is neatly filed in a labeled manila envelope and consists of hand-colored graphic worksheets, handwritten source code, music and sound tables, advertising copy, game packaging, a receipt book, and the sales brochures and advertising he prepared under his business names, Supersoft Software and later Fun Division. Such an archive is rare in 1980s homebrew game production and a significant resource for the homebrew historian. It is ironic that a homebrew developer with modest resources should have kept better documentation than many commercial game companies with larger budgets, but it is yet another example of the polemic/aesthetic/ethical significance of everyday practice: "To appropriate information for oneself, to put it in a series" and "to defend the autonomy of what comes from one's own personality"(de Certeau and Giard 1998a, 255). In the next section, I drill down into Marentes's practice and examine some of the archival documents surrounding the production of his TRS-80 and Tandy Color Computer games before focusing on the documentary materials for the game he considers his best, *Donut Dilemma* (1984).

Marentes was very ambitious, both creatively and entrepreneurially; in our conversation about his 1980s practice, the two are always linked. He freely admits that the reason he initially wanted to make original games was to pursue a dream of becoming a multimillionaire videogame designer, selling millions of copies of games around the world, in order to retire. Marentes's emphasis on creating original games was based on the perception that if he was to make money from marketing his games, legally they needed to be his creations. As he explained, "That way when I do break into the billionaire league I've got a product that no one can sue me over."

Professional Approach

His comparative youth notwithstanding, Marentes aspired to a professionalism in the development and marketing of games that he felt was missing from his immediate context. He pinpoints 1982 as the beginning of his "commercial programming effort[s]," though he had been "writing many games in BASIC since my parents bought me my first computer" (Marentes n.d.d). Marentes's "first" game was a BASIC and machine language hybrid, sold through "user groups and small adverts in local TRS-80 magazines." With the next game, *Cosmic Bomber* (1982), he "felt better prepared to tackle a full assembly language game project." He explains, "It is quite a difficult language to learn; it involves a lot more steps than what BASIC does. I always hated the fact that anything I wrote in BASIC, I couldn't get that same quality that you could get in the arcades. Programming in [assembly] language felt like you were actually playing with the electronics, the hardware. You actually did manipulate the way the process—the electronics—worked."

Marentes approached a Brisbane tobacconist who was also a TRS-80 machine language programmer. The shop owner had an assistant selling tobacco while he sat out the back of the shop and programmed. Marentes had noticed that he had TRS-80 software on the shelf, so he approached him with *Cosmic Bomber* and asked whether he'd be interested in selling it. Though initially dubious, the shop owner agreed when he saw that the game was in machine language, as Marentes tells the story. The game apparently didn't sell well and has been lost.

The same year (1982), at the age of eighteen, Marentes registered a company name in Australia—Supersoft Software—intending to market his own games. Marentes's aim was to create software on a par with what he considered to be the best internationally. As he put it, "My goal was to get my programs sold commercially in Australia, step one." He considered it part of his market research to play others' games, explaining that:

> In any competitive environment you have to look at your opposition, so if I was going to be writing games I had to look to see what the opposition was creating and I had to make sure that my games were at least as good as, or better than, theirs. . . . Personally I do think that I created software that was better than them. I was aiming my software to be on par with a software company that was

in the US, known as Big Five Software . . . they were doing all the really good quality stuff.

In 1983, Marentes made a Letraset catalog of all the games he and an associate, Richard Lindley, had written up to that point (figure 4.4) and sent it to all his purchasers, using the Supersoft name. The black and white master for the catalog not only summarizes each of their games, showing screenshots, but on the reverse side also describes the pair's software as "QUALITY SOFTWARE FOR THE GAMER DESIGNED BY GAMERS." They make an appeal for the quality of their software, presumably at least in part to justify its reasonably high asking price ($16.95 and $19.95):

> The price of computer game software these days is expensive but what is worse is that their quality is no better. There are many game programs on the market selling for over twenty dollars that only use trivial graphic manipulation routines and even simpler sound synthesis routines. Our software is priced according to the quality of the program. Our latest games, *Moon Scout* and *The Gladiator*, clearly show what we mean. More sound and graphic effects for your dollar. But good graphics and sound is only half the story. A unique and challenging game concept such as our program *Neutroid* is just as important. And adventure games such as our *Stellar Odyssey* (and now under development, *Stellar Odyssey Part II*) have about 50% of their development time devoted to story concept, 25% on graphics and sound development and the remainder to programming. . . . In short, we believe our software to be excellent value for money and guaranteed to provide the gamer with many hours of enjoyment.
>
> But we don't bother raving about our software, we let our customers do it for us!

After discovering that there is a company in England with the same business name, Marentes changed his to a name he hadn't seen before, Fun Division. He designed a logo and took out an advertisement in *Your Computer* magazine for his 1984 game *Donut Dilemma*—the game he considers his best for the TRS-80—all while keeping up a steady output of game titles (three per year in 1983 and 1984). The ad for *Donut Dilemma* is seen in figure 4.5.

The inspiration for *Donut Dilemma* came from the donut kiosk that Marentes's family owned. He explains, "Occasionally, something would go wrong, usually in the part that flips the donuts over, and the donuts would get all messed up. Seeing this one day, a revelation hit me. Why not do a game based on a donut factory where everything has gone wrong?'" He made his own platform, with ten levels, including the "fat splurter" and "icing sugar" levels.

Marentes's archive for *Donut Dilemma* provides a glimpse into his working method and influences. Born in 1964, Marentes was around twenty at the

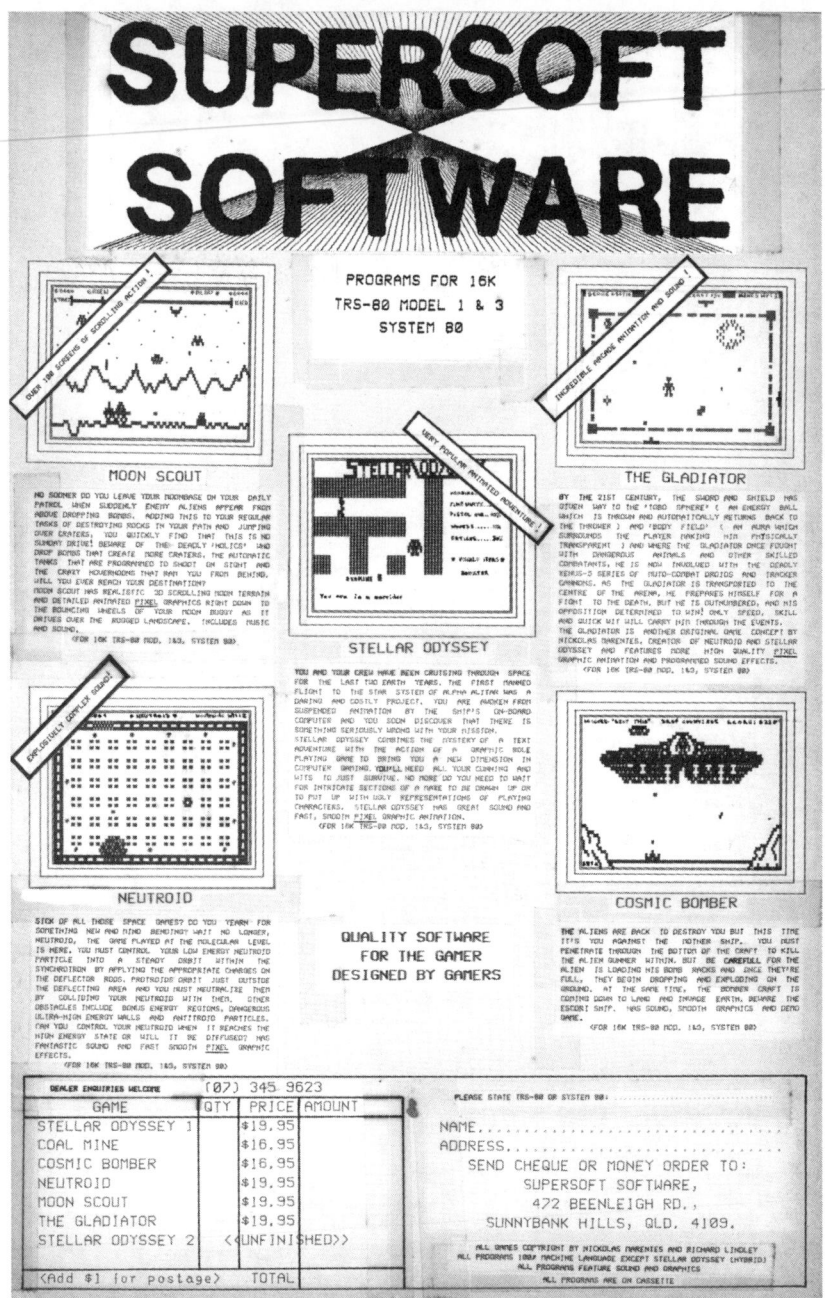

Figure 4.4
Supersoft Software sales brochure. Courtesy of Nickolas Marentes. (See plate 8.)

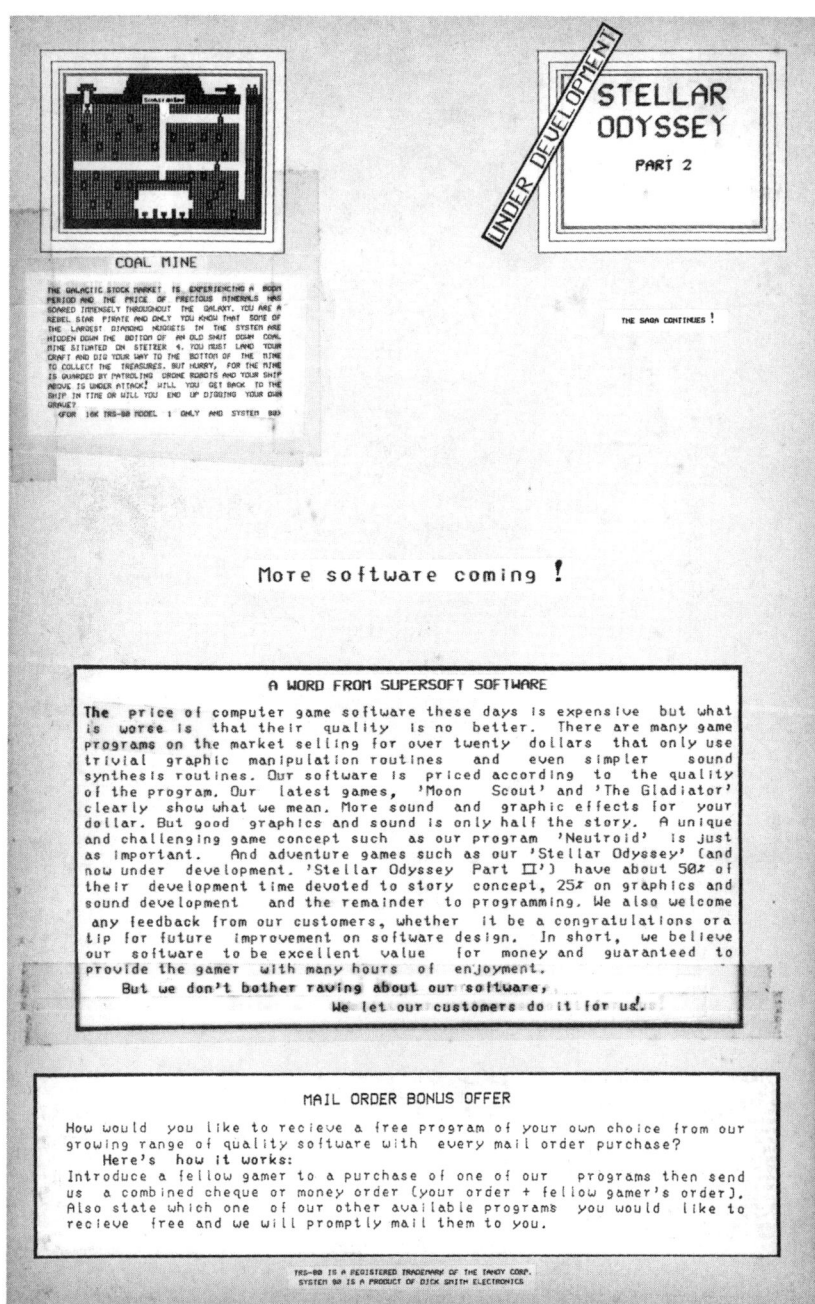

Figure 4.4
(continued)

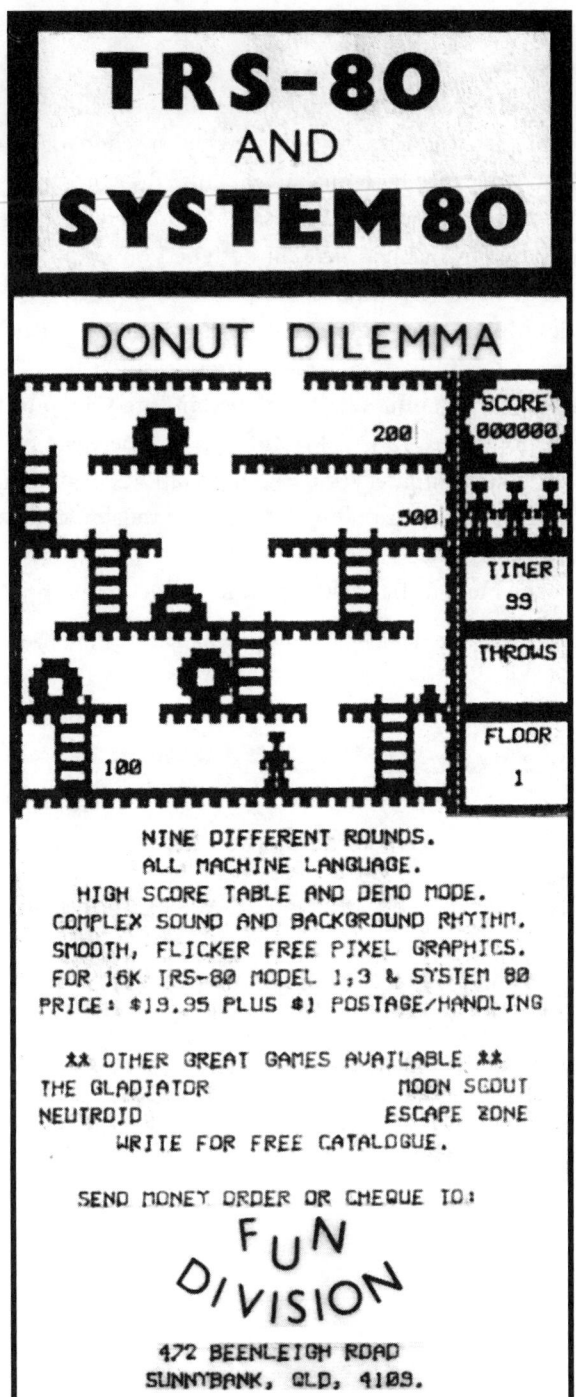

Figure 4.5
Ad for *Donut Dilemma* published in *Your Computer* magazine. Courtesy of Nickolas Marentes.

time of the game's release, so he had by then finished school. Nevertheless, he continued the technique he developed while at school of writing source code longhand on paper before typing it into the computer. Marentes's code contained no comments, yet he says, "I could read this easily." The graphics for Marentes's games were all designed on graph paper, as seen in the sample in figure 4.6. This homespun technique of designing computer graphics—literally, by coloring in squares on graph paper—renders computer graphics as something strange, or at least less familiar than they are today. The blocky graphics that they became are shown in the sample Level 2 screenshot (figure 4.7). As Marentes says, "There was no such thing as software on your computer for designing graphics or designing games. Back then there was none of that like there is nowadays. So it was all done [with] pencil and paper."

Marentes explained his handwritten source code to me as follows:

> The graphics were stored in memory in the computer . . . so I had to document where I stored it all, so they were the memory locations and that was like the

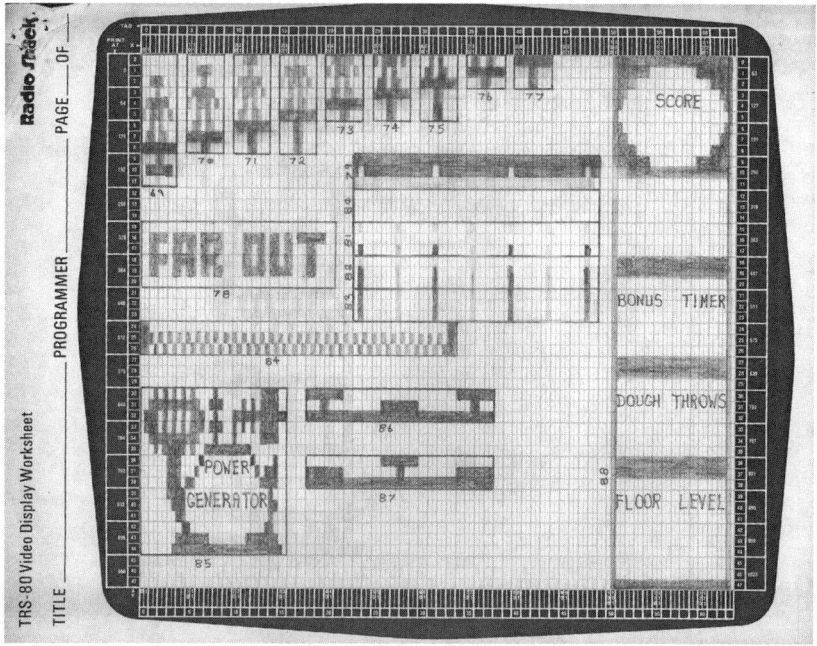

Figure 4.6
Graph paper graphics for a level in *Donut Dilemma*. Courtesy of Nickolas Marentes. (See plate 9.)

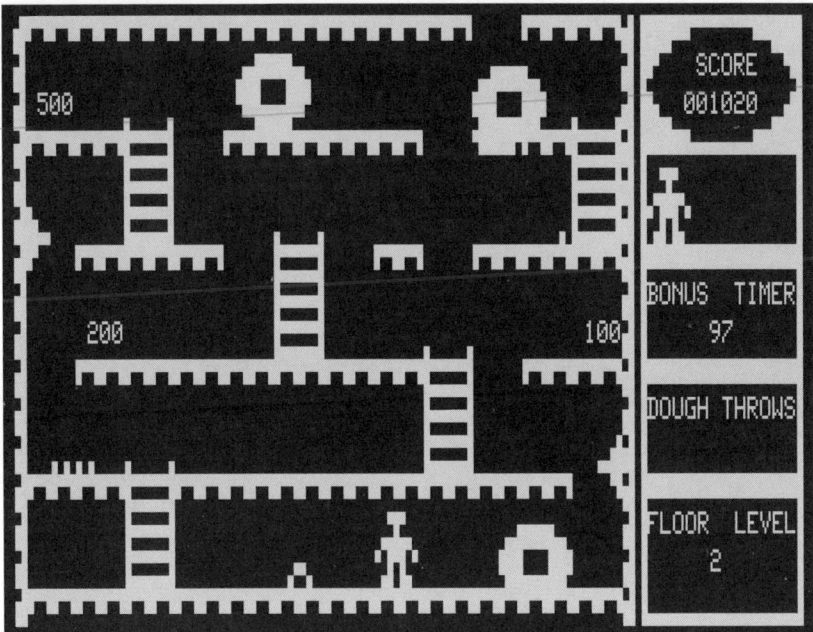

Figure 4.7
Donut Dilemma Level 2 screenshot. Courtesy of Nickolas Marentes.

graphics number. I numbered a graphic block . . . number 1 was there. I'd have to look at it all [again], but it's all data telling me where the graphics are and how it was laid out in memory so when I'm coding it I knew how to draw it from there and put it up on the screen. Same with the sound. I actually created sound and I stored the sound effects that I created in certain locations, and then I needed to know where they were so they can play it in the code.

The music and sound table shown "in all [their] graphite glory" in figure 4.8 tells not only the locations where sound effects were stored but also the clear inspiration of 1980s pop music acts Devo and Yazoo, including Marentes's use of one of the most recognizable synthesizer riffs of the decade, from the opening bars of Yazoo's "Don't Go" (1982). Elsewhere I have argued that sketches of artwork on graph paper for 1980s computer games convey what it might have been like to create graphics for an 8-bit computer game in a way that playing the game does not (Swalwell 2017a). Something similar holds for the soundtrack. That early 1980s synthesizer riff—remediated through the TRS-80's speaker—evocatively locates *Donut Dilemma*.

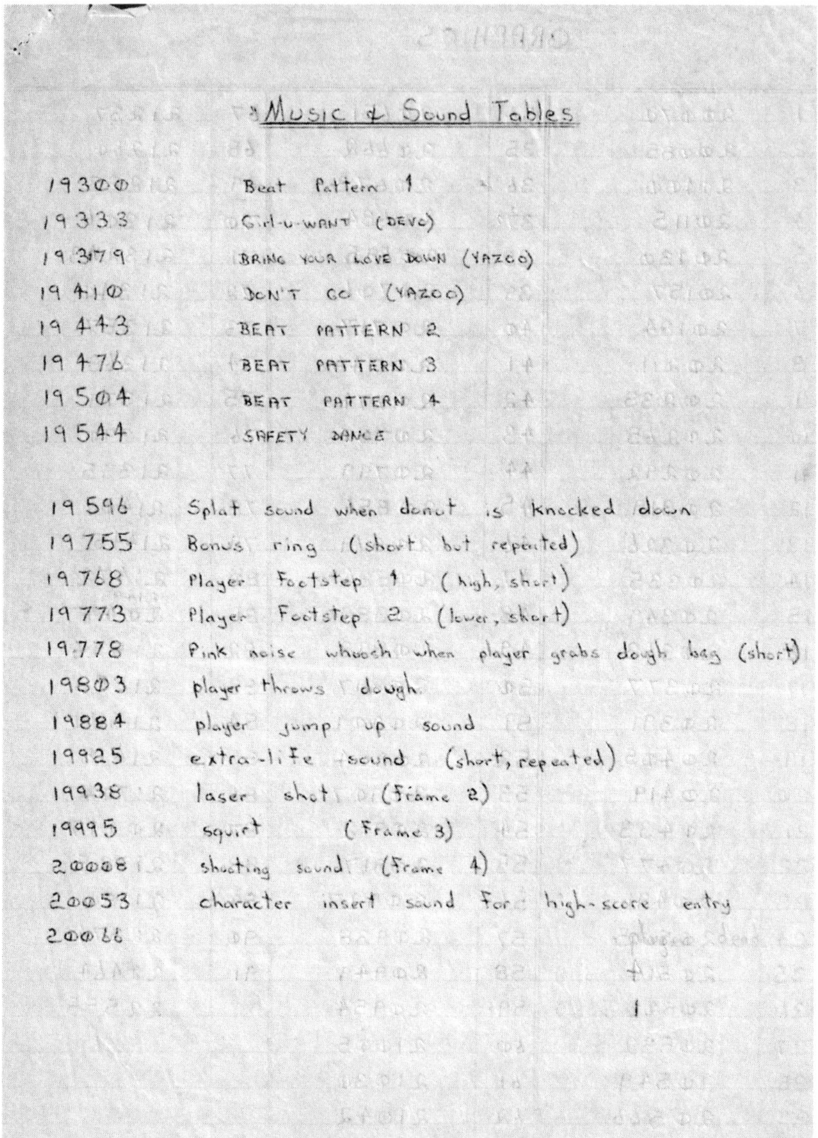

Music & Sound Tables

19300	Beat Pattern 1
19333	Gid-u-want (DEVO)
19379	Bring your love down (YAZOO)
19410	Don't go (YAZOO)
19443	Beat Pattern 2
19476	Beat Pattern 3
19504	Beat Pattern 4
19544	Safety Dance
19596	Splat sound when donut is knocked down
19755	Bonus ring (short but repeated)
19768	Player Footstep 1 (high, short)
19773	Player Footstep 2 (lower, short)
19778	Pink noise whoosh when player grabs dough bag (short)
19803	player throws dough
19884	player jump up sound
19925	extra-life sound (short, repeated)
19938	laser shot (frame 2)
19995	squirt (Frame 3)
20008	shooting sound (Frame 4)
20053	character insert sound for high-score entry
20066	

Figure 4.8
Music and sound tables for *Donut Dilemma* (showing musical inspiration from Devo and Yazoo). Courtesy of Nickolas Marentes.

Once a game was written, Marentes would develop packaging, which comprised printed instructions and artwork. Figure 4.9 shows the packaging and instructions for *Donut Dilemma*. Marentes explains:

> What I used to do back then with the lack of a desktop publisher, everything was done on paper. I just printed everything out on paper, cut, glued, maybe hand drew a few things up, and made up these master templates which have all faded now and the sticky tape's gone yellow. But back in the day I would then go to the local photocopy place and say, "Okay, run me off 20 copies of this." You'd fold that in half so that was the front cover, the instructions were inside and then the back cover would just be the blurb about the game. And then that would wrap up into a plastic bag and then the cassette would be written out on my computer and I'd package the cassette tape in the bag, and that was a product.

Marketing

On marketing *Donut Dilemma*, Marentes reflects:

> If I could have marketed the game properly in the US via a big distributor like Adventure International or even Big Five Software, I believe it could have wiped the table in sales. But alas, I was just a "small fry" operating on a limited budget (nothing) far away from where the real action was, so again, all my sales were restricted to club meetings and catalogue post-outs to past customers. I felt that this game had so much potential so I did create a small paying ad into a major computer magazine. This got me a few more sales and to date, *Donut Dilemma* was my best selling TRS-80 Model 1 game. (Marentes n.d.a)

After developing a couple of games in 1984—including one for the Tandy Color Computer, known affectionately as the CoCo—Marentes's game development activity seems to have had a bit of a lull. Then, deciding he wanted a game that Tandy would market in all their Australian stores, he ported the TRS-80 Model 1 version of *Donut Dilemma* to the CoCo and sent it to Tandy Australia's head office. He writes, "I kept my expectations low. In the past, all games except for a few educational titles were imported from Tandy Corporation's main warehouse in the US. That seemed to be where most of the decisions were made as to what became the product line" (Marentes n.d.b).

Tandy Australia liked *Donut Dilemma*, and Marentes shipped the first order for 1,000 packages in August 1987 (Marentes n.d.b). According to Marentes,

> [Tandy] just said, "Look, we want to sell your game. We'll buy this many copies. This is what you've got to do. You've got to have them delivered by then." Bang, that's it. And, you know, you send the box of all the goods and then they distribute it out to 350 stores in Australia. I didn't make a lot of money on it because the deal was,

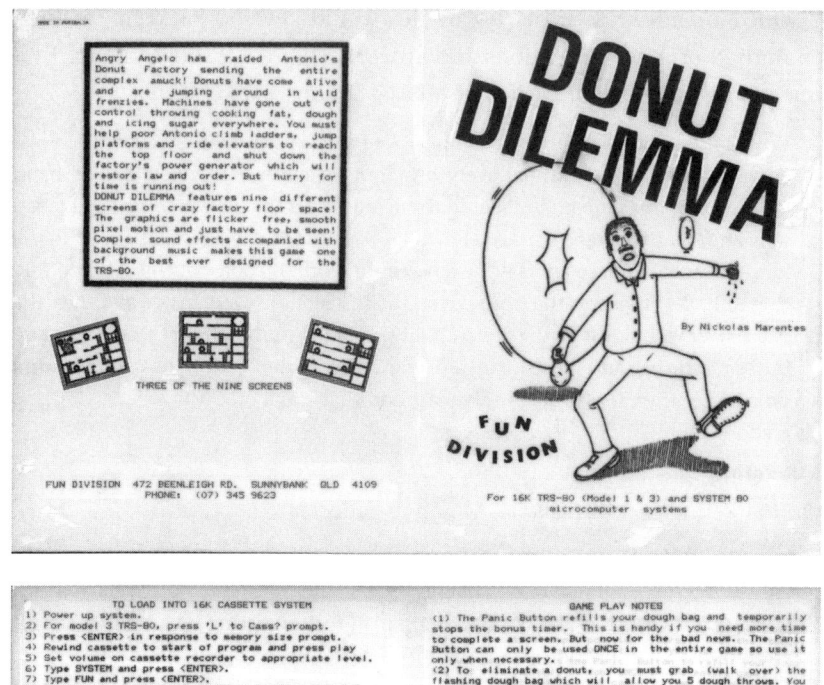
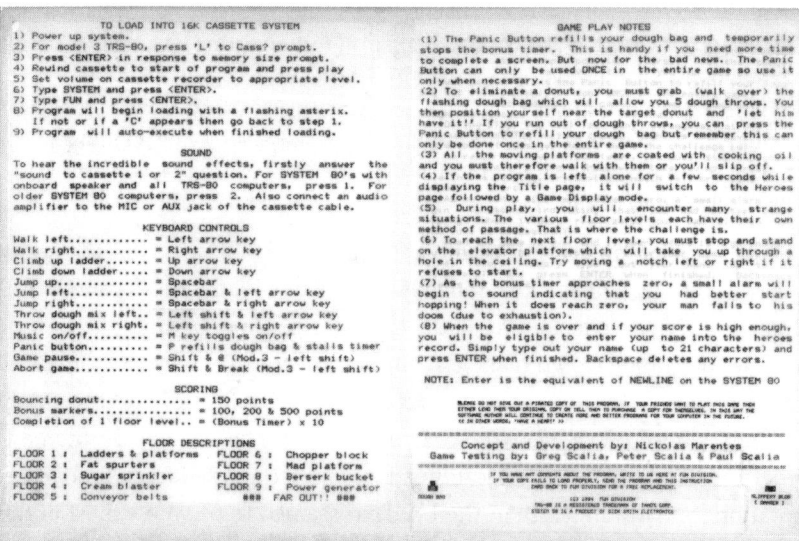

Figure 4.9
Donut Dilemma packaging, including instructions. Courtesy of Nickolas Marentes.

I think they paid $6 per tape; they sold them for $10 each, of course, but I only got $6, and that included my costs. So really, at the end of the day I probably made $3, $3.50, but $3.50 multiplied by 3,000 to a kid who has just finished school seemed like a lot of money, and it was the first step to becoming that billionaire.

There were some differences between the CoCo version and the earlier TRS-80 game. The CoCo version focuses on sound effects rather than music. Marentes also had a job by this stage, so he could afford to commission some artwork for the cover. It is a sketch of his character Antonio. He stipulated that the character had to look "ethnic" because his father is Greek, though it ended up looking more Italian than Greek, he thinks. Color photocopying was also now within reach, so the artwork was more colorful than previous covers. Then, he recalled, "later on when I got the big order for 3,000 I thought, well, I need something better than a colour photocopy, so I actually went to a printer." The cover art was complete with the Tandy logo and the product's assigned catalog number, but the biggest difference is that the game itself is in color. It looks great. In total, Marentes's *Donut Dilemma* on the CoCo sold 3,500 copies through Tandy Electronics, thanks partly to its being bundled with the Color Computer 3 in a Christmas package. Marentes would go on to have two more games marketed by Tandy Australia: the first, *Rupert Rythym* [sic] (1988), sold 850 copies, but the second, *Space Intruders* (1988), only sold 300 copies. Marentes attributes the declining sales to Tandy's "understandable" promotion of its new range of IBM-compatible computers. He noted that "the rest of the market had moved on" to 16-bit systems such as Commodore Amigas, Atari STs, Apple Macs, and IBM 286 PCs (Marentes n.d.c).

"Produce[d] without Capitalizing"

Marentes's resolve to monetize his games sets him apart from most homebrew game developers, or at least those of a similar age. The majority of younger homebrew authors viewed their activities as a fun hobby and "produce[d] without capitalizing" (de Certeau 1984, xx),[10] though many programmers seized opportunities to make some money out of their hobby or otherwise turn some advantage from it. Some (Passfield, Clarkson, and Perry) sold their games to publishers, either outright or for royalties, while Symons derived income from his books. Though this did not net them large sums of money, as teenagers they were well pleased with their royalty checks—except when they bounced—and with the fact that their games or books were being published. Publication provided a form of recognition and esteem. Meanwhile, Sibly exchanged his game *Dinky Kong* (1984) with an Auckland computer store owner for a floppy disk drive, enabling him to save programs to disk.

In seizing opportunities to make some money out of their hobby, homebrew creators were operating in the context of a discourse that emphasized just this: the chance to get games published and make money from their hobby. In encouraging people to contribute their programs, magazines talked up the possibility of hobbyists turning professional. For example, the editor of *Sega Computer* wrote encouragingly, "It is important to note that this is YOUR magazine. So please send in any programs that you have, be them [sic] small or large, complex or simple . . . it matters not. If someone sends in a program and someone else learns from it then it has been worth it! To be quite frank I could name ten people in the UK and three in New Zealand who now make a lot of money through writing programs, and they all started by writing a few simple programs and having them published in computer magazines! SO GET WRITING!" (Anonymous editor 1986).

Rewards from homebrew game development existed on a continuum from those who (might have) sold one game right through to those who set up small businesses. Of the three business owners in my sample (Arthur Streeter, Darryll Reynolds, and Nickolas Marentes), two were older. The possibility of youngsters hitting the big time was, however, a prominent motif, with obvious appeal. We see it in a story headlined "Class of 1982," which profiled a number of industrious teenagers who—ahead of their planned university studies in electrical engineering and computer science—were making money writing programs for computers. Michael Fackerell, Garry Epps, and Martin Foord were Dynamic Software. Fackerell explained, "It happens quite often . . . that computers are released before they have software ready." The trio had experience writing programs for the Apple, TRS and System 80, Compucolour, and the Microbee, and were considering getting a Vic machine. "We've now got about eight programs for the Microbee. . . . We're working all the time. They're mostly games at the moment," Fackerell added (Filatoff 1983, 27).

Shops and software houses often placed ads in magazines inviting programmers to submit their programs for possible publication. An ad placed by the well-known Australasian chain store Dick Smith Electronics in the December 1983/January 1984 issue of New Zealand magazine *Bits and Bytes* advocated that "enterprising computer buffs" profit from their hobby by writing programs for the new Dick Smith Colour Computer (Dick Smith Electronics 1984). That same month, suburban Perth company Mytek Computing entreated readers to send in their MicroBee programs for review by using an eye-catching "Wanted"-style ad asking "let us show you the benefits of working with our team of design, production and distribution specialists" (figure 4.10). The advertisement in question was a cheeky rip-off of a Broderbund ad (figure 4.11), indicating that soliciting software through magazines was also done in the US market.[11]

While profiting from one's hobby was part of the discourse surrounding programming of micros in the home in the 1980s, there is a degree of instability or slippage around this apparent hobbyist mercantilism. While teenagers were understandably excited to earn some money from their programming, the returns were usually quite modest. Some aspired to sell their games and were not able to. For instance, after having the code for his game *Harbour* published in *Computer Input* and selling *City Lander* to Grandstand for $300 in the same year, John Perry wrote another game that he also showed to the company. Grandstand, however, was not interested, as in *Dungeons beneath Cairo* they already had a title that was similar—and considered superior—to the one Perry was offering because it was written in machine code rather than BASIC. Others, such as Andrew Kerr, who intended to try to have his untitled machine-coded game for the Sega SC3000 endorsed by Poseidon Software, never got around to it ("but it did get me an 'A' in Computers at college!," he said). The narrative of prosperity from doing something you love often turned out to be less amazing than people had perhaps hoped, as Marentes's story illustrates.

Cottage Industries: Marentes, Streeter, Reynolds

Ultimately, Marentes didn't break through and realize his ambition to ride the game development wave to prosperity, though he certainly tried. He did eventually get three of his games distributed in the US via the distributor Game Point Software, and these were profiled in the magazine dedicated to the Tandy Color Computer, *The Rainbow* (1989). Ironically, Marentes's intimate knowledge of and dedication to programming the Tandy computers may have been one reason why he did not realize his ambition. Marentes invokes another of my informants—fellow Queenslander John Passfield,

Figure 4.10
Mytek rip-off of Broderbund advertisement, highlighting the entrepreneurial message that was put out to homebrew software developers. *Your Computer*, vol. 3, no. 5, December 1983, p. 126. Collection of the State Library of NSW.

Figure 4.11
Broderbund Software "Wanted" advertisement. *Antic: The Atari Resource*, vol. 1, no. 3, August 1982, p. 6. Courtesy of James Capparell, publisher.

whose output he is now aware of—offering that "he [Passfield] was smart; as soon as the Amiga came out and there was a bit of a game industry coming up there he went to that and started developing for that." By contrast, Marentes reflects on his own twin motivations:

> I wanted to be a billionaire but it was also a hobby. I actually liked the machine and I liked writing for that machine. So I guess I was tethered that way too much; I should have actually said, "Okay, I've got a lot of ideas on how I could do better, but I should forget about those and move on to the next computer where there's an upcoming market now. This computer's been around three, four years, is approaching the end of life . . . and end of life is not a good sign." Everyone else

was upgrading to the new machines. I should have upgraded as well. Whereas, like a heretic, I just stuck to it because it was still a hobby for me as well. You know, I like coding and I like the challenge of finding out how to make it do more and make it do things that it couldn't do. And I got stuck in that.

Exactly what "end of life" means requires detailed consideration, which I will get to shortly. Before I do that, I will contrast Marentes's experience with those of two other informants—Streeter and Reynolds—who managed to turn their homebrew activities into quite lucrative enterprises.

Value for money, distribution, and conversions set Streeter's and Reynolds's apparently commercially successful homebrew game development businesses apart. The challenge for anyone trying to sell game software in this era seems to have been to convince people to exchange their cash for a product whose quality they weren't sure of, as Streeter's account that follows attests. While Marentes's games were very experimental (e.g., *Neutroid*), and I gather technically very good, the prices he was asking were significantly higher than others at the time. Like Marentes, Arthur Streeter's local reach was supplemented by the postal system, but Streeter's prices were significantly lower and he packaged multiple games on one tape, thereby apparently providing better value for money. Streeter got started when the Vic-20 came out, as he explains:

> The first colour computer under $300. . . . Of course I couldn't resist that. I had to have one of those! And my kids at that stage, David, my son, was old enough to want to be involved with playing games, and I found that games were quite expensive and so I started to write my own. And then, of course, I used to give copies to a friend who had a VIC-20, and he said to me one day, "Why don't you market them?" And I thought, "Oh, I don't know about that." I thought, "I'll invest $10 in this," and I bought I think four tapes for $4 and [spent] $6 for an ad in the local newspaper and got a good response. So it sort of built up from there. I'm not a risk-taker, and had the investment required been more like $50 I probably wouldn't have done it. But since it had virtually no cost to me to try it, I tried it, it seemed to work, and then I just kept writing games and selling them through mail order. So I put ads in newspapers and people would ring up, give me their name and address and I'd post them out a catalogue and they'd then send the . . . I had a tear-off form to order what they wanted and then I'd post it back to them.

Streeter first advertised his games in the classifieds of the local papers, then in the Sydney-based newspapers, the *Daily Telegraph* and the *Herald*, which were distributed across the state. Streeter said "it went remarkably well." He attributes his success to "a little original thought":

> The original thought was that a tape, the difference in cost between a 10 minute tape and a 90 minute tape is small, so why not put more than one game on the tape? And so I used to typically put 10 games on the tape, and that way if I sold the tape for $10 people thought or people felt they were getting reasonable value, whereas a commercial game would have cost perhaps, I don't know, $15, $20, something like that at that time. So I was giving 10 games for less than one game; admittedly they perhaps weren't as, what's the word for it, they weren't as elaborate, [they were] simpler games. But for a certain age group they were very well received, and I did sell a lot.

Streeter may not have been a businessman, but he had a keen vernacular understanding of where people might look—and therefore where to place ads people might respond to—to purchase games. He also understood that giving people more games for a reasonable purchase price was likely to be popular and that the overhead of copying tapes with a high volume of sales was very low. Streeter recalls that he did sell single titles, "but I think I charged something like $3 or $4 for a single game; consequently most people bought the compilation game."[12] With these insights into ordinary culture, he managed to turn his hobbyist homebrew practice into a rewarding entrepreneurial sideline, supplementing his salaried position as a microbiologist. As mentioned in chapter 3, the income he earned from Street Games allowed him to take other risks in his working life, offering him the ability to experiment (and "trace [his] own path") with an idea he had for software that would be useful in the hospital lab environment.

While selling games was a profitable sideline for Streeter—one from which he earned "more than pocket money"—Darryll Reynolds was a businessman who came to love computing and started a successful homebrew game programming business, first under the name Gameworx and then as Softgold. A self-taught programmer, Reynolds's initial programming efforts in the early 1980s saw him write software for the TRS-80 to run the accounts of the business he then had, building playground equipment. He quickly "got hooked on computing." Games were what he was interested in, so when the Vic-20 came out, he explained,

> I got that and I thought "well this could mean, shall we say, a mass market piece of hardware." Shops were springing up, not that many but they were springing up, retail computer shops, and they were selling software and hardware and I thought "well this is an opportunity." So I thought "well why don't we try it?" I put together a number of games, packages to sell, did all the artwork, bags, you know we'd copy those, we'd staple the bags together at night, and I'd take one or

two days off during the week and run around to the few computer retail outlets that were here in Victoria and direct sell. And I sold a lot. I sold a hell of a lot.

It seems that Reynolds's distribution deals were what gave his one-man operation such a considerable reach. He continues:

> While I was doing this I ran into a company called Computer Classics. They were . . . at the time Video Classics and they were a marketing company, selling videos to video rental stores. They saw the opportunity again with . . . computer software, and they set up a company called Computer Classics, and they approached me. . . . To cut a long story short they . . . said "We're prepared to offer you a contract of X amount of dollars, a quantity of so much. You sit down and write the titles and we'll do the packaging and sell from there."

For a time, Reynolds successfully capitalized on the different types of microcomputers on the market. He continues, "Then the Commodore 64 appeared on the market, which was so much better of course, because it had the extra memory and it was a lot better system. I converted what I'd done at that stage across to the Commodore 64. And well that's basically how I got into it from there. And then, as the years went on, there were a number of [other] steps as to where I evolved . . . and the other systems I got onto."

After Computer Classics got out of the market, Reynolds was approached by one of their marketing people, who "set up a company called DotSoft to do exactly the same thing." Reynolds explains,

> He was right into chain stores. [The supermarkets] Coles and Woolworths were selling software then, [and the department store] Myer. They were selling it everywhere. And DotSoft really did a lot of work marketing. . . . They approached me and said "Can you do this? Can you do that?" Then they said "Can you convert across to this other platform?," Sega, for example. We got all the manuals, got all the hardware from Sega, all for nothing, and worked out how to drive that machine and then converted our titles into a package that was suitable for Sega. I think there was a [system] called the MSX system, a couple of others [such as] Oric, and eventually we got onto the Amstrad which was being handled by AWA, who have since gone under, of course, a number of years ago. I worked with them . . . doing my packaging and marketing, once DotSoft sort of disappeared. I was doing everything myself. I was doing the artwork, getting it printed, packaging it up and taking an order from AWA, for example, or somebody else. And they would buy and distribute it out to their agents and sell it from there. So I was really selling to a wholesaler I suppose.

When I asked Reynolds whether it was fair to characterize his business as a cottage industry, he responded:

> Yeah, absolutely, but it turned out to be more than just a cottage industry I'd say, [because] I was selling overseas. . . . I thought "well, I'm selling this here through these other companies locally, what about overseas?" And I looked around . . . and I found a couple of companies [including] Severn Software in England. . . . We did a lot together and he marketed a lot of my software under a different name. The first major one I did out here was *Secret of Bastow Manor*, and we marketed that very, very similarly as *Mystery of Munroe Manor* over in the UK and through Europe. We did it all like that and I was just getting royalties coming back.

Reynolds estimated that if average weekly earnings were between $120 and $150 per week, then he was easily earning $700–$800 per week. It came in "fits and starts," but it was "multiples of the average wage at that stage." While homebrew game development existed on a continuum, Streeter and Reynolds are clearly on the edge of my homebrew definition, with Reynolds's experience challenging the criterion of small-scale, local distribution.

Reynolds pivoted his business around 1987, getting out of game development and into writing business software. He cites a range of reasons for this decision: "Piracy started to cut his [Severn Software's] business to pieces. And the big budget titles were coming along and sort of overtaking our market and [Severn] just couldn't compete with the advertising that they were putting out. And at that stage people . . . they wanted the big name titles: the thing that they'd seen advertised in magazines. I got out of the industry—the games industry—around about 1987 I would guess. So, it was about six years I was working there."

At roughly the same time as Reynolds decided to get out of game development, several other factors were making themselves felt. These would change the landscape for homebrew as well as for home microcomputer use more generally.

Game Over?

By 1985, as already mentioned, 8-bit computers were starting to give way to 16-bit machines. Of particular note in Australia and New Zealand were the Amiga computer, released in 1985, and the Amstrad, the distribution of which Grandstand New Zealand took on in early 1986. New consoles were also released, such as the Super Nintendo Entertainment System and the Sega Mega Drive. Homebrew developers were affected by the new generation of platforms. According to Stephen Cass (2014), the consoles presented less of a threat to general-purpose home computers than did the newer computers and the consolidation in the market that the increasing dominance of PCs and Macs (released 1984) heralded, because consoles were not easily programmable. Nevertheless, budgets for games were getting bigger, and production values were increasing. Several of my informants made the switch to writing software for other machines, whether the Amstrad (Reynolds), IBM (Streeter), or the Amiga (Passfield and Sibly), the latter much admired for its multimedia capabilities. Marentes was impressed by the Amiga—as he says, "[I] got one . . . back then to be my more powerful machine for doing development work on"—but he never made the switch to programming for it, or indeed any other non-Tandy platform.

Aside from the new generation of computers, there were other changes afoot in both the computer and game industries. The fortunes of the company that made Microbee computers went south just after the decade's midpoint, with it posting a $970,000 loss in the six months ended December 31, 1986; those like Clarkson, who had published games through its software arm, Honeysoft, found that their royalty checks bounced. Though the company passed through several sets of hands in an effort to keep it afloat—first the laser printer distributor Impact in late 1987 (West 1987) and the following year computer entrepreneur Giuseppe De Simone, who set an Australian Stock Exchange record for the cheapest takeover of a main board-listed company, buying Microbee for $133,000[13]—the company eventually folded in the early 1990s under pressure from PC clones. Meanwhile, in New Zealand, Grandstand ceased its support for the Sega SC3000 in early 1986, as it began distributing the Amstrad. I wrote in chapter 2 of how the magazine *Sega Computer* passed through a series of hands before senior school student Michael Hadrup took it on. Though it was clearly no longer profitable for a company to continue producing the magazine, Hadrup's valiant effort indicates that people were still using their Sega computers at least until 1988, when the SC3000 was five years old.

For Streeter, the changed structural conditions in the computer and computer game markets also presented an opportunity to—like Reynolds—pivot his interest. As he says, "I was pretty much a solo operator, and so I guess I came, I saw, I wrote games [laughs], and then I kind of disappeared. Well, perhaps my interests moved in a different direction." Streeter made the leap from writing games for 8-bit platforms into nongame software development. By the early 1990s, he relates,

The Games

> I . . . gave away my salaried position for starting my own software firm . . . [creating] auto-analyser database applications for hospital environments, so for laboratories, x-ray departments, cardiology departments, wherever they have a lot of data and typically machines of one form or another. We used PC-based networks to link on the one side to auto-analysers, taking the output from machines of one form or another, and on the other side to the hospital's mainframe computers which were ward-based systems. . . . And so we were kind of a link between the people doing things in the back rooms as it were—doing tests of mostly diagnostic stuff of one form or another. . . . And then, when they'd sort of massaged the data to their satisfaction on a particular test or a particular patient, then our system would output it to the hospital mainframe where it was accessible at the ward level.

That Streeter saw the need for such a software application was the result not only of his highly specialized biomedical background and years of experience working in hospital labs but also his experience writing games in BASIC. I was surprised to learn that the programs developed by Synchrotech Software (known as *Lab Link*, *Image Link*, and *DVATS*) were based on a program that he originally wrote in QuickBASIC: "I wrote the original version and then after we got going I had other programmers enhancing it and so on." He explained, "I know about that feeling about BASIC being a beginner's language, after all the B in BASIC stands for Beginner. But the thing is, don't base it on why it came into existence; base your assessment on what it can do. And effectively what happened is computers got so fast that although BASIC wasn't a particularly quick language—that was its major drawback, the execution speed was limited—the hardware speed just meant that became irrelevant . . . our programs used to run very satisfactorily on 286 computers." As already mentioned, Streeter emphatically credits the income he made from developing and selling games with giving him the freedom to branch out, take a risk, and try something new.

Mainstreaming Personal Computing

From the middle of the 1980s on, changes were afoot that would begin to shift the public's perceptions of computers and play a part in computers becoming more widespread. Recall that in chapter 2 I cited Eric Lindsay's memorable phrase that "a computer is not a toaster" as he bemoaned low uptake numbers for microcomputers, which I tied to perceptions of computing's usefulness (or lack thereof). As the 1980s wore on, more people who were not hobbyists began to embrace home computing. I will detail three aspects of this shift.

First, computers were increasingly presented as a communication tool, with magazines profiling the coming "communications revolution." Australian Telecom's Viatel—"the national videotex service"—began operations on February 28, 1985. Paul Zabrs began writing a semiregular column on the basics of "computer communication" the following month for the *Australian Apple Review* (Zabrs 1985). BBS listings became more frequent in magazines.

Second, users no longer had to know how the computer worked. Journalist and publisher of several Australian computer magazines Gareth Powell writes:

> In most communications sessions using a personal computer, you really don't have to know how the modem or the software works.
>
> There is no operator's test as there is in ham radio.
>
> You truly do not have to be born with a soldering iron in your hand to be able to communicate with a computer. (Powell 1986, 46)

As a pragmatic user himself, Powell noted that he was less interested in how it all worked than in the fact that it *did* work. Powell's stable of magazines all attempted to explain computing from a beginner's point of view. Very aware of the size of the untapped market, they tried to guide the reader through what they needed to know in as nontechnical a manner as possible. Though the magazines' production values were not high, the editors managed to successfully convey the idea that computing was accessible to those new to it, with columns such as "New to Computing?" and "Especially for Beginners" regular features in many of Powell's titles (e.g., Farrell 1985, 2). In August 1986, Powell launched the non-brand-specific magazine *Australian Home Computer GEM*—GEM standing for games, entertainment, and music. Magazines such as *GEM*, together with the *Australian Commodore and Amiga Review Annual*, represented a new type of computer magazine. No longer were they just addressing those interested in the computer as a programming platform; instead, they were at pains to point out that in its new guise as the locus for various forms of entertainment, the computer was for everyone. But it is important to note that right up until at least the turn of the decade, magazines continued to publish instructional articles on different aspects of programming—if no longer actual code listings—for those interested. For instance, the final scanned issue of the *Australian Commodore and Amiga Review* that the Internet Archive holds (the December 1989 issue) features articles on using sprites on the Commodore 64, learning Amiga Basic, Z80 machine language on the Commodore 128, programming in C on the Amiga *and* Commodore 64/128, an Amiga assembly language tutorial, and plenty of game reviews and entertainment news. Entertainment computing did not kill off home programming, and indeed these magazines demonstrate that the invitation to program as a hobby was extended to Australian users of 8- and 16-bit computers simultaneously.

Third, by the middle of the decade, a range of new, software-based user activities were being discussed. Increasingly, the music, graphics, and art capabilities of computers were highlighted (Richardson 1985). Powell wrote an article on desktop video editing, detailing what he did to the opening scenes of the movie *Psycho* on a computer (Powell 1985). At the turn of the decade, desktop video was being talked about in Amiga magazines. The computer became associated with entertainment, with the computer as "software player" (Haddon cited in Veraart 2011, 57).

Desktop or personal publishing arrived as one such software-based activity. Such "productivity software," as it was billed, allowed home users to produce personalized stationery; publish newsletters; make certificates, invitations, and banners; and print out just about anything they wanted to, or so the marketing spiel went. You could create cartoons with Garfield and begin to manipulate photographs with graphics utility programs such as *Cockroach Graphics Utility*. Software packages offering these functions received glowing reviews and were credited with generating "an enormous upsurge of interest in programs that actually 'do something,' particularly with a printer hooked up" (Anonymous 1988). Ironically, it seemed that paper's materiality trumped all: the front cover of the *Australian Commodore Review* in May 1987 featured programs for mechanical toys to print out and assemble. This range of programs boosted the perception that a computer in the home was indeed a useful purchase. By 1990, computing was more graphical, more mouse driven (complete with discourses on "intuitive" interfaces), and, in the eyes of the purchasing public, more useful.

Decline Theses

The changes I have just narrated relating to the rise of home computing during the 1980s are not particularly controversial: the user base expanded and more-powerful machines and a greater array of software became available. Nevertheless, these changes have generated some interesting assertions in game and computer history circles, specifically around how usage changed following the early microcomputing moment. Thus far, I have been referring to early microcomputing using a reasonably broad date range, beginning in the late 1970s and continuing through the 1980s. This is partly because it can be hard to pin down dates and out of a recognition that things happened at different times in different locales. However, I now need to attend more closely to the temporality of practice in order to critically engage with the work of two scholars—Graeme Kirkpatrick and Frank Veraart—who have identified particular points of inflection in microcomputer history.

Kirkpatrick and Veraart have studied microcomputer use and gaming in the UK and the Netherlands, respectively. Both make useful contributions to the scholarship on early computer uptake. In several articles and a book, Kirkpatrick presents arguments based on content and discourse analyses of two computer game magazines from the UK, *Computers and Video Games* and *Commodore User* (Kirkpatrick 2015, 2014, 2012), while Veraart presents an account of a Dutch computer club (Veraart 2011). Kirkpatrick's and Veraart's accounts are very different—they use different methods and are informed by quite different theories—yet they both make very specific claims about how and when computer use changed in the 1980s. Veraart's and Kirkpatrick's accounts are simultaneously interesting and puzzling to me because they are so at odds with my findings. Specifically, both allege a decline in programming activity in either the early part of the 1980s (Veraart) or halfway through it (Kirkpatrick). I will briefly reprise their arguments before laying out my reservations and critique.

Veraart's article "Losing Meanings: Computer Games in Dutch Domestic Use, 1975–2000" (Veraart 2011) claims to "[show] how games in domestic use lost their versatile meanings beyond entertainment" via a study of the Hobby Computer Club (HCC) in the Netherlands, which would become one of the largest clubs in the world, with 68,000 members by 1988. Veraart writes that the HCC was established in 1977 as a "user group of computer enthusiast peers, similar to most hobby computer clubs in other countries. However, by the mid 1980s, HCC started addressing a more general public and changed into a much broader organization on computing" (Veraart 2011, 52).

Given the club's early establishment, its first members were understandably interested in the technical aspects of computing. Veraart writes, "In May 1978, one year after its founding, it had 767 members, and by September 1979, the club had 2,500 members with an interest both in computer building and programming. About half of the members that joined the first year had homebrew computers or kits" (56). Veraart attributes a hacker ethic to these hobbyist members: they were interested in games, which were seen as "acts of craftsmanship . . . not merely fun applications, rather they served learning and exploring purposes" (56).

From such beginnings, Veraart details the club's expansion and the appearance of newer computers, which he claims changed how members used their computers: "In the early 1980s new computer types appeared. In contrast to homebrew and kit computers, hobby computers changed to electronics encased in a box. Of these the Tandy Radio Shack-80 (TRS-80), Commodore's PET 2001, Exidy Sorcerer, and Apple II became the most popular among Dutch hobbyists. Hobbyists with an interest in programming, rather than the machines' electronics, started to dominate computer clubs in the late 1970s" (56).

Veraart goes on to present a thesis whereby users began tinkering with electronics and then turned to programming the computer before turning away from programming in the early 1980s to utilize the computer as a "software player"—claiming that "hobbyists shifted their attention from making programs to using them" (57) with the introduction of the Commodore 64, Sinclair ZX80, and Phillips P2000—and then to cracking and hacking (illegal copying). This claim is presented against the backdrop of the HCC's concern with its image. According to Veraart, the club was embarrassed by the cracking and copying activity and wanted to be viewed as "a serious partner in personal computer developments" rather than as a "gaming society" (61).

Kirkpatrick has argued over the course of several publications that there was a discursive bifurcation between computer hobbyists and the emergent category of "gamers" during the mid-1980s. On the basis of content and discourse analyses of magazines, Kirkpatrick concludes that 1985 was the key year in which there was a repudiation of computing in favor of gaming in the UK (Kirkpatrick 2015, 2012). Code disappears from the magazines in his sample, and these adopt different emphases and a range of evaluative criteria specific to games, distancing themselves from "computer nerds" in the process. "Gaming," Kirkpatrick writes, "had to secure its autonomy from computing" (Kirkpatrick 2012). Gaming becomes strongly identified with playing games, and previous associations with writing games fall away in these magazines. Kirkpatrick goes on to claim that hobbyists abandoned coding their own games in the middle of the decade as hobbyist computing parted company with gaming, in part because the former was deemed "uncool." Noticeable across both accounts is the broadening of the computing base and an increasing emphasis on games as entertainment, which I am not arguing against. It is the claim that earlier types of experimental hobbyist engagement with computers were supplanted and that people simply stopped programming games on their micros that I take issue with.

I find the decline theses in both these accounts curious because, as stated, I have not found evidence of such a decline in hobbyist or homebrew practice in either Australia or New Zealand, and certainly nothing that would indicate a change or bifurcation of activity as early as the release of the ZX80 which came out in the UK in 1980 or in 1985.[14] In fact, as detailed, the *Australian Commodore and Amiga Review* was actively working both ends of the market; that is, new users and technically adept users. Changes were certainly afoot in the mid-1980s, but I would explain them differently. Just as people came to homebrew via a range of different routes, their trajectories and development also varied. This occurred against a backdrop of structural changes that saw some who were sensitive to market forces change direction (e.g., Reynolds and Streeter pivoting their businesses). But, as hobbyists, most of my other informants' motives were intrinsic, so while their practice understandably changed as they grew and matured, their reasons for doing what they did were invariably complex and personal—more than one-dimensional. Even Nickolas Marentes, who had harbored the ambition of being a software billionaire, eventually let his business ambition go but continued to program as a hobby, "for the love of coding."

A word about methods is in order because we need to think beyond whether people actually traded in one practice (programming) for another (cracking and hacking, according to Veraart) to ask how we can know whether they did or not. And, as valuable as archival research is, Veraart's account confuses representation and technical possibility with practice. Though microcomputers became capable of acting as software players, that doesn't mean they ceased to be used also for tinkering or programming. Uncovering use or practice requires different methods, which can then complement archival research. This is particularly the case when the practice in question is a hidden or private one, as with homebrew. Although he claims to be taking a "user perspective" approach, Veraart's account is very hardware driven and teleological, based exclusively on archival study. The evidence he cites does not support his conclusions about practice: the claim about progression is entirely unsupported. I do not doubt that cracking and hacking went on and would have concerned the club's powers-that-be, as Veraart notes, but it is not convincing to draw inferences about other practices being supplanted without offering evidence. There's a need for circumspection in the conclusions drawn from archival research alone.

One possible explanation for Veraart's findings in the HCC archives pertains to different types of hobbyists. By his own account, different members joined over time. Club members are effectively treated as though they all had the same motivations and did the same things with their computers. Distinctions between the curious layperson who might have seen the computer as a software player and those who had more technical understanding and derived pleasure from tinkering or programming are flattened. Veraart's framework cannot account for their coexistence. By contrast, Tom Lean usefully discusses two groups or audiences who were involved as computer hobbyists, "mass market users" on the one hand and hobbyists on the other (Lean 2016, 65), with some computers, such as the ZX80, crossing over between these groups.[15]

With regard to Kirkpatrick's thesis, while I find much to admire in his scholarship, and some aspects of his account (and Veraart's) resonate with what my informants have told me,[16] our findings are significantly different. While I have observed the effort to broaden the base of computing in the address of magazines such as *GEM*, I have not discerned the split between computing and games that Kirkpatrick does, and I dispute his claim that people stopped programming in 1985. Some of the points of difference can be explained by the fact that the UK was a much larger market than the ones I have studied. It is likely that magazines were able to specialize to a much greater degree, dedicating themselves to particular segments of the market, such as those interested in playing games rather than making them. This would seem to be a matter of editorial policy rather than use per se, so the decline thesis Kirkpatrick advances—that people stopped programming in 1985—is still questionable. Moreover, the archival evidence he summons *cannot* prove that homebrew development ceased. Once again, being an activity conducted in private domestic space, the lack of visibility of programming in these magazines is hardly surprising, especially given that the titles in question are constructing their target audience in opposition to such practices. Nor is it particularly surprising that my study should arrive at different conclusions, given the different methodologies employed.

I have been establishing how people used microcomputers, how they came to know how to program them, and some of the experimental coding activities they undertook, which ultimately resulted in games and other software being produced. I have done this through recourse to interviews with users who reported on their activities, and I have combined this with archival evidence. While ethnography is usually taken to be a contemporary pursuit involving first-person observation, perhaps an expansion in the use of this term is warranted for a study such as this, despite my interviews with informants having occurred three decades after the fact. (I noted in chapter 1 that Giard referred to de Certeau's study of "possession" in seventeenth-century France as "a pioneering ethnography"; three decades is temporally much closer than the three centuries in de Certeau's study). Though some might query the role and reliability of memory in oral history interviews, many of my informants' claims—particularly about when things happened—are verifiable by reference to the dates that software was published, and I have sought to cross-reference these wherever possible. This is one strength of a mixed-methods approach to computer history. Had either Veraart's or Kirkpatrick's study involved research with users themselves, I suspect they would have had to amend their arguments about the decline in homebrew programming practice significantly. They would also have had to nuance the rather patronizing attitude that is implicit in suggestions that audiences simply adopt—or cease to practice—certain activities and attitudes based on what they read in media or on what other club members do, without regard for their own interests, predilections, and circumstances. While practice is shaped by discourse such as that found in magazines, it is not determined by it.

Programming Continued

While the middle of the 1980s brought a number of changes, hobbyists in Australia and New Zealand didn't cease programming. The peak in homebrew production in Australia and New Zealand seems to have been from the late 1970s to the mid-1980s, but the invitation to code continued to be extended, variously through user groups, magazines, and schools, alongside invitations to make other uses of a micro. A late 1986 issue of the magazine *Australian Apple Review* (vol. 3, no.10) presents an assortment of uses, including playing computer games (there are reviews of *Dambusters*, *Boulder Dash*, and *The Hobbit*); reviews of "computerised playfulness" (a review of *The Great International Paper Airplane Construction Kit*, dubbed "sheer genius") as well as desktop publishing and video animation programs; setting up and running a bulletin board; pimping one's Apple with reviews of a modem and floppy disk drive and ads for a range of expansion cards; artistic use of a "Macinitizer," featuring a pencil-like stylus; and—significantly—part four of a series on Apple assembly language.

Several of my informants produced titles for their 8-bit computers in the late 1980s and even into the 1990s. They were clearly still motivated to continue developing games. Matthew Hall's *Jewels of Sancara Island* was programmed in Turbo Pascal in 1988 for the Microbee computer. Hall wrote it for an eighth grade computer class at Edenhope High School in Western Victoria, and it is "the only [early game of his] that anyone else ever played" (Hall 2013). Hall's creation of this game in 1988 demonstrates that schools continued to use the installed base of Microbees well after the company's financial difficulties started in 1985. Dorothy Millard was another who kept developing adventure games. She repeatedly told us she got great satisfaction from doing it and this, together with the contact she had with those who would phone her for solutions, illustrates that homebrew game development continued to deliver intense rewards and satisfaction for her. Millard's last game listed on Gamebase64 is dated 1994, and she explains in an interview that this was when she "re-entered the workforce and simply didn't get around to learning a new program when I switched to the Amiga" (Gunness 1999), demonstrating that homebrew games continued to be created for the Commodore 64 well after its distribution ceased.

Technologically focused accounts can easily miss the texture around technology's adoption, and the longevity of its use. Just as practices did not supplant one another in neat fashion, hardware transitions were also messy and quite drawn out. For instance, new forms of external storage became available in the mid-1980s, but adoption was neither linear, quick, nor uniform. As I detailed in chapter 2, in the Australian and New Zealand markets, computers were expensive, middle-class discretionary purchases, saved for over long periods. There were many models to choose from, and purchases were often thoroughly researched. Even though floppy disk technology was available in some early 1980s computer systems (e.g., IBM, Apple II, but also the TRS-80), it was often relatively pricey and therefore seen as an unjustifiable expense. By contrast, magnetic tape technology was inexpensive. Nickolas Marentes explained to me that although he was developing his games using floppy disk storage, games continued to be published on tape because it was the standard; everyone had one:

I was selling software, so in any sort of a market you have to cater for the lowest common denominator. See, for example, if I write any software I had to make sure that software can run on the maximum number of computers in order to maximise the sales. So hence, I had to supply the software on cassette tape, and it was a requirement back then, because I sold those programmes through Tandy. Tandy actually wanted the software on cassette, to sell on cassette. They didn't want it on disk because they knew that most of the Color Computers out there were still cassette-based; it was a home market. Different if you're talking a business computer because they all had disk drives, but a little home computer, it was still a lot of cassettes.

Technical availability does not equal uptake. Just as some Sega SC3000 owners in 1988 were still using their machines five years after they'd been released, Marentes was still outputting games onto tape for sale as late as 1988, which indicates how slowly that first generation of microcomputer hardware was replaced. Tape was important in Australian and New Zealand markets and was used for a long time. People didn't upgrade straightaway; they still don't. And even when they did upgrade and the requirement to type in source code was removed, they didn't stop writing code, as seen in Sibly's acquisition of a more sophisticated disk-based storage system, gained through bartering a game he'd programmed with a computer store owner. Sibly wanted to do more than just play other people's games. He wanted to write his own. In charting transitions in microcomputer use and practice, we are talking about tendencies, not hard delineations.

Teamwork

While some hobbyists were happy to keep developing in a low-key way for personal satisfaction or for local distribution and consumption, rising production values from the mid-1980s meant that it became harder to write games for the commercial market as a solo operator. In New Zealand, Sibly teamed up with friends Cameron McKechnie, Rodney Smith, and Blair Zuppicich to develop the Amiga game *Sorceror's Apprentice* (1990) under the name Art Software. McKechnie, Smith, and Zuppicich also released *Sirius 7*, a horizontally scrolling, space-themed shoot-'em-up the same year. The credits for *Sirius 7* demonstrate the different roles that were starting to characterize game development by the end of the decade, with design (Smith), programming (McKechnie), and music (Zuppicich) each getting separate mentions (de Vries et al. 2013). There is a Let's Play of the game on YouTube, and the music receives a special mention in the comments, with xcimbal writing: "Unusually nice music" ("SIRIUS 7 [AMIGA—FULL GAME]—YouTube" 2013).

Meanwhile, in Australia, John Passfield—whose creation of *Chilly Willy* as a schoolboy had given him such satisfaction—found himself a somewhat disillusioned computer science graduate in the early 1990s. He hadn't done much programming since the decline of the Microbee. After a stint working in a dead-end job for the telephone utility company Telecom—during which time he didn't even own a computer—Passfield bought an Amiga and learned to program it, teaming up with Steve Stamatiadis to set up Interactive Binary Illusions. They remade his 1985 Microbee version of *Halloween Harry* (de Vries et al. 2013). But even then, as Passfield recalls,

> I didn't really have ... even when we began what was turned into Interactive Binary Illusions in Brisbane ... I remember calling a company in Sydney who was a distributor of games and sort of getting a feel of "what's the market like?" and "can we make money?" And the guy said, "Oh, you might make a few thousand dollars." So it really was a passion thing. And in my head I kind of knew that that didn't seem right because I knew that there were people in the US who were making games and were probably multi-millionaires, so it didn't quite add up. But I didn't quite get it, like it felt like, "Well, that's too far away. We're in Australia and there's no way we'll make money here."

Halloween Harry appeared for the PC in 1993 (also known as *Alien Carnage*). *Flight of the Amazon Queen* followed in 1995 for the Amiga.

The mid-1980s brought a number of changes, among them new and more powerful computers, a growing user base, and rising production values in the game business. The effects on homebrew developers were mixed: homebrew authors were not immune to factors such as newer, more powerful platforms and higher production standards, but there were plenty of people who continued to code beyond the point at which their systems ceased being contemporary. There was no coup de grâce because homebrew developers were not solely driven by market values and business logic. As such, the arguments for a definitive watershed year are not compelling. Pinpointing a particular moment and arguing that is when previous activity was eclipsed ignores too much contrary evidence. In chapter 6, I will ponder the significance of those who continue to code for 8-bit computers today, well into their computer's fourth decade, arguing that such practice enables us to grasp the historicity of digitality. But before jumping ahead to the present, I have one more important set of practices associated with 1980s microcomputing to unpack: hardware hacking.

5 Hardware Hacking and Electronics

The question of user productivity has long been of interest to theorists of screen and media audiences, as I discussed in chapter 3. The foundations of much of this scholarship lie in literary studies and film or screen studies. Reading has long been privileged as the figure for audiences engaging in cultural production, whether via the idea that readers are also writers or the cultural studies concept that a viewer may undertake various readings of a text as part of an interpretive or hermeneutic endeavor. The case of homebrew game development—and indeed the larger question of what users did with computers in the 1980s—raises some profound questions regarding the adequacy of the reading metaphor, for while it is intelligible to speak of users *authoring* code and *writing* games, 1980s micro user practices went well beyond the realm of writing. Some users were also building computers, hacking hardware together, and troubleshooting problems with their microcomputers' electronics, practices that are not adequately captured under the banner of "reading." In this chapter, I outline these practices, consider why we have heard relatively little about them to date, and argue that we need other figures to capture user engagements with technology, particularly electronics and hardware.

Build Your Own

User invention and experimentation in the micro era were not confined to software creation. I have reviewed an extensive range of Australian and New Zealand primary source materials pitched at the hobbyist sector, including computer magazines, electronics magazines, hackers' handbooks, "circuit cookbooks," instructional "build your own videogame" style books, and information on locally made kit computers. Reviewing these materials, it is clear that people built electronics as part of their early microcomputer use. While it is hard to ascertain just how many people were engaged with electronics, audited circulation figures give some indication: crossover computer and electronics monthly *Your Computer: Magazine for Business and Pleasure* was selling just under 20,000 copies per month by 1983 (Audit Bureau of Circulations 1983).[1]

I was surprised by just how extensive the archival traces of electronics engagement were in 1980s microcomputer publications. Many magazines I consulted presented circuit diagrams, a mode of address of users I had not been expecting to find. Years earlier, I had come across two beautiful hand-assembled keyboards on a visit to the store of the Berlin Computerspiele Museum (see figures 5.1 and 5.2). Understanding that these were from the former East Germany, I had reasoned that they owed their existence to the scarcity of computer components in the old East.[2] People couldn't get their hands on commercial peripherals, so they had to "make do" in the Fiskean sense, building their own. The then director, Andreas Lange, explained that the COCOM embargo restricted technology transfer to prevent its use for

Figure 5.1
Homemade keyboard, collection of the Berlin Computerspiele Museum. Photo by Melanie Swalwell. (See plate 10.)

Figure 5.2
Another homemade keyboard, collection of the Berlin Computerspiele Museum. Photo by Melanie Swalwell. (See plate 11.)

military purposes (Švelch 2018, 13). But as I trawled through published materials in the Mitchell Library in Sydney, I found projects to build in Australian magazines that reminded me of those keyboards, which did not owe their existence to difficulties with supply. I have come to realize that hobbyist electronics and hardware hacking is another overlooked aspect of early microcomputer practice, not noticed by media or computer historians because we weren't looking for it and not thought worth mentioning by those who were involved because to them it was unremarkable. As I was noticing the incidence of electronics and engineering viewpoints in microcomputer culture, snippets from interviews I had conducted with some of my New Zealand informants came back to me. I had also previously read of the Combined Microcomputer Users Group—an informal network of Auckland clubs and groups—and their initiative to build low-cost acoustic modems for members (Arrow 1985, 110). Following these realizations, asking whether people were involved with the electronics side of their computing became one of my standard interview questions.

Computer hardware provided users with many opportunities for "fiddling around." A number of early hobbyist microcomputers came in electronic kit form, requiring that users first assemble them. Jamieson Rowe's EDUC-8 is the earliest Australian "do-it-yourself computer," with the instructions appearing in serialized form in the magazine *Electronics Australia* beginning in August 1974 (Rowe 1974). Though Rowe's design was initially thought to be the first such kit microcomputer published anywhere, it was narrowly beaten by a *Radio Electronics* publication of the design for the Mark-8 computer just weeks earlier. Other kit computers available to build in Australia included the DREAM and the Microbee from Applied Technology, the Super 80 (Tanton n.d.), and the Applix 1616, which debuted as the *Electronics Today International* project 1616 (Morton and Berger 1986; Lindsay 1989). The late 1970s and early 1980s was the cusp of when computers came fully built and when it was still possible to build them from components, sometimes purchased in kit form. The Australian 8-bit computer, the Microbee, was initially offered for purchase in kit form, and was featured on the cover of *Your Computer* magazine and in a thirty-two-page supplement inside in February 1982.

Existing electronics hobbyists were some of the first adopters of computers. They *needed* to understand microprocessors like they had needed to understand radio. New Zealanders Neil Breen and Selwyn Arrow both built their own computers early on. Breen worked as a programmer for the (little known) New Zealand office of the (very well known) arcade game manufacturer Taito. He recalls, "I was building my first computer as an amateur on veroboard in 1976. I built several machines for myself. My wife was running the local Plunket membership lists on a Z80-based machine with probably about 16k of RAM in the late 1970s."

Paralleling those who wrote games and sold them, Breen also built computers to sell as a "sideline" when he was working for Taito. When I asked Arrow about his first computer, he explained that he'd "started to build one in the late 1970s, just out of bits and pieces, which is what you had to do in those days. I got the keyboard done." Arrow recalls he had gotten the impetus from reading *Byte* magazine:

> It was either Christmas '77 or '78, more likely 1978. . . . A copy of *Byte* magazine arrived. . . . I read it twice, including all the ads. It just opened up a whole new world. . . .
>
> I had decided that I would start with peripherals and then eventually we'd sort things out. I was planning on using sockets and connectors and things that were surplus at work, you see, old bits, to make cards that plugged in. And eventually I

realised there was this S100 interface, which meant that it was a socket with 100 pins. Eventually I dropped that one.... But in the meantime I was studying a book on microprocessors—the Z80—which of course was in the [first computer he bought, the Exidy] Sorcerer. So I wasn't wasting my time totally. I had this wonderfully large—by today's standards—keyboard with all the bells and whistles on it. I never really used it in warfare.... I took it apart eventually.

Arrow's account of his entry into the world of microcomputers—driven by curiosity and excitement, encouraged to try out ideas, and prepared to build them (mainly to see how they worked, not to use them)—is emblematic of what I call "the will to mod." Fiddling around with electronics and computers seemed pointless to many people at the time, who struggled with the apparent uselessness of computing as a hobby. Yet, in electronics there was no shame in undertaking a project for curiosity's sake. Indeed, an ethic of playfulness and enthusiastic curiosity was central in hobbyist electronics circles. Such an ethic is evident in the naming of groups like the Brisbane PC1500 Bit Fiddlers Club and in the advice of those who urged users to take things apart to see how they worked.

Home hobbyists had begun to build and mod electronic game consoles in the 1970s, prior to the popular takeoff of microcomputers. A range of how-to electronics guides and schematics invited readers to build their own consoles, such as the Selecta-Game shown in figure 5.3. Several book titles unpacked the intricacies of TV game devices. Len Buckwalter's *Video Games* demonstrated how cellophane screen overlays could be used to "make your own game" (Buckwalter 1977). Two years later, Walter Buchsbaum and Robert Mauro's quite extraordinary electronic engineering and hobbyist how-to guide appeared, called *Electronic Games: Design, Programming, and Troubleshooting*. In it, Buchsbaum and Mauro break down the different elements of electronics involved in such games, claiming that "engineers, students, technicians, and competent hobbyists who already know electronics will find in this book all the information they need to design, program, maintain, and troubleshoot all types of electronic games" (Buchsbaum and Mauro 1979, ix). Richard Giles—an informant on a related research project—dabbled in building various electronics projects as an adolescent in Adelaide in the 1970s, with a "ping pong game" among them. He had grown up "making things and making music—light organs, colour organs, things like that" before finding plans to build a game console "in *Electronics Australia* or somewhere.... It would have been [made from] integrated circuits, simple digital integrated circuits without any computing power at all." Like Arrow,

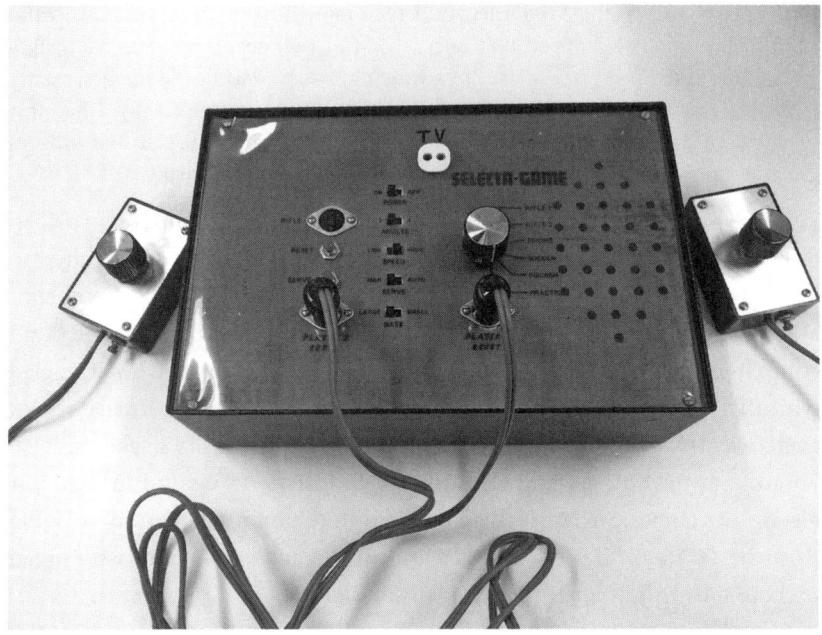

Figure 5.3
Selecta-Game, homemade console. Computer Archaeology Lab collection. Gift of Philip Kocent. Photo by Denise de Vries. (See plate 12.)

the interesting part for Giles was, as he put it, "mostly the making . . . we got it to work [and then] moved on to whatever was the next thing."

While Giles knew a couple of people who were also making things—he ran a small business with another boy at school for a time "set[ting] up lights for discos and parties" and his father was involved in electrical work—he remarked that electronics was "a lone sport really." This resonates with Arrow's account of his retreat into a corner of the lounge room for two years, which was how long it took him to learn about the computer and undertake the voluntary positions he held with the New Zealand Microcomputer Club. Arrow was keen to tell me that his wife was very understanding of his newfound fascination with computers and even shared in it to some degree. She had grown up in a family of car enthusiasts and so "was used to becoming involved in projects," as he put it. As the child of a car enthusiast myself, I recognize something of a shared do-it-yourself drive between auto and computer tinkering: it is partly a need to be working with one's hands technically, putting skills to some good use. There's often an element of thrift involved, through avoiding waste and effecting repairs economically. Sometimes, the curiosity that drives the will to mod is more accurately described as a compulsion: an insatiable curiosity and desire to have a go.

Ham Radio Antecedents

This period of building games and tinkering with other electronics projects in the 1970s was a precursor to the programming of micros that would start to appear only a few years later. Giles narrowly missed the micro era, having already programmed a PDP-11 in 1974 at university as part of his electrical engineering degree. But the sort of knowledge that amateur electronics hobbyists developed, not to mention the concentration and self-study skills required of practitioners, would certainly come in handy in the microcomputer era. In hindsight, hobby electronics prepared the ground—as ham radio had done before it—for those who would embrace tinkering with microcomputer hardware.[3] Microcomputing inherited parts of its culture from electronics and ham radio, not least because many micro hobbyists (including Arrow) were themselves hams. Ham infrastructure was also sometimes leveraged: R. Harrison notes that a great many of the first Australian computer hobbyists were licensed radio amateurs and recounts the first meeting of MEG—the Microcomputer Enthusiasts Group—on January 17, 1977, at the then NSW Wireless Institute Centre (the radio amateurs' organization headquarters) in Sydney (R. Harrison 1985, 9).[4] And it was not uncommon to find authors publishing their call signs in electronics magazines after their name. Indeed, local technical writers—some of whom were active during the period, such as Jamieson Rowe (Rowe 1974)—continue to publish their amateur radio call signs in their biographies (Silicon Chip Publications n.d.).[5] Perhaps not surprisingly, it wasn't long before attempts to transmit microcomputer programs over short wave frequencies started to appear in computer magazines. Shayne Doyle reported in the December 1983 issue of *Bits and Bytes* on several such experiments in New Zealand, among them that "Denis Young (ZL2BFI) of Raumati South and Jim Wilkinson (ZL2WI) of Waikanae were able to transfer programs [between their Microbees] reliably at both 300 baud and 1200 baud, in spite of some initial difficulty with the signal from the MicroBees being distorted by the ICOM 22S transceiver microphone amplifier" (Doyle 1983, 86).[6]

Hobby electronics of this sort is a form of what Robert A. Stebbins has called "serious leisure." Stebbins classifies "makers and tinkerers" as one of the major subtypes of hobbyist. Of serious leisure, he writes: "It is profound, long-lasting, and invariably based on substantial skills, knowledge, or experience, if not on a combination of these three. It also requires perseverance to a greater or lesser degree. In the course of gaining and expressing these acquisitions as well as searching for the special rewards this leisure can offer, amateurs, hobbyists, and volunteers get the sense that they are pursuing a career, not unlike the ones pursued in the more evolved, high-level occupations" (Stebbins 2001, 54).

Stebbins's reflections on the satisfaction that serious leisure delivers—"the steady pursuit of an amateur, hobbyist, or career volunteer activity that captivates its participants with its complexity and many challenges"—are reminiscent of both Giard's observations of the "profound pleasure [of] practicing a modest inventiveness" (de Certeau, Giard, and Mayol 1998, 213) and of her and de Certeau's articulation of the aesthetic, ethical, and polemical rewards of everyday practices (de Certeau and Giard 1998a, 254–255).

But it was the keen culture of experimentation that most clearly connects earlier electronics and radio hobbyists with micro hobbyists. Users were experimenting—in Williams's sense of trying and testing (Swalwell 2008a)—with what it was possible to do and to create. Whether they were programming or hacking hardware together, they largely did it "for the technical challenge and thrills" (Haring 2007).

Hacking and "Circuit Cookbooks"

The experimental and curiosity-driven engagement with electronics is found in many published accounts of people building computers, including the unpredictability of the results. Eric Lindsay bought one of the earliest Microbee computer kits. After much labor, he had to send the kit back to the manufacturer. "They spent about four days on it before also giving up. The offending board was returned to the supplier as an example of problems, and with a new board my MicroBee started running and has been trouble-free since," Lindsay reports (Lindsay and Moffat 1982). By contrast, Moffat's initial construction was trouble-free, but:

> within an hour of completion, the first problem surfaced: heat, and lots of it. . . . There is an old rule of thumb in the electronics business, . . . if a part is too hot to touch, it's too hot! Just about every active part of the power supply produced painful burns. Inquiries to Applied Technology brought the response that "all the parts were running within their ratings." [My friend] J. J. and I, being of

a more conservative nature, found a source of Sinclair ZX81 plugpacks rated at 9.5 V/1.2 A. . . . Another Hobart MicroBee user didn't get there in time—his 12 V plugpack "blew its guts." (Lindsay and Moffat 1982)

Not everyone built their own computer, of course. Microbees, for instance, though initially available as kits, were later available for purchase fully assembled. But there were still opportunities to become involved in hardware assembly or modification, for instance with peripherals. Sometimes, there was a collective dimension to the activity, as with the previously mentioned Auckland project allowing for the manufacture of low-cost modems (Arrow 1985, 110).

Many how-to projects on modifying a computer were published in the pages of Australian electronics and some computer magazines in the 1980s. These included attaching peripherals and other hardware interventions to satisfy those who might desire a hardcopy printout cheaply or a joystick for playing games, or who wished to overclock their computer or any one of a large number of other possible "enhancements," for which circuit diagrams and instructions were typically provided. The magazine *ETI (Electronics Today International)* also published what were known as "circuit cookbooks." These supplements collected "interesting and useful circuit and design ideas" from the magazine, together with others that hadn't appeared owing to space constraints (Roger Harrison 1985a). That guides on how to experiment and tinker with electronics and micros were analogized as cookbooks resonates with the argument I advanced in chapter 3 that—as far as experimentation is concerned—programming a computer and cooking are analogous activities. While recipes outline ingredients and a method with some precision, when an ingredient or utensil is lacking, there is only one thing to do: improvise. Then, as Giard notes, "the recipe itself loses significance, becoming little more than an occasion for a free invention by analogy or association of ideas, through a subtle game of substitutions, abandonment, additions, and borrowings" (de Certeau, Giard, and Mayol 1998, 201). The hardware equivalents of experimentation would seem to be building, modification, disassembly, troubleshooting, and repair. In the words of our esteemed Microbee kit builder Eric Lindsay, "After you have some experience making simple projects, you will want to design your own or make modifications to existing equipment" (Lindsay 1983, 114).

How-to projects and circuit cookbooks encouraged and normalized a tinkerer's—or, as it was then known, a hacker's—ethic in early computer culture (R. Harrison 1985). The term *hacking* was frequently used to describe the bringing together of various items of hardware, typically in the service of extending a computer's capabilities. The Microbee, in particular, was actively marketed in terms of its ability to be modified and "hacked." By the end of 1983, Applied Technology was extolling the benefits of the Series 2 model—actually called the "Experimenter"—with a remarkable advertisement featuring a robot arm interfaced with a Microbee pouring a cup of tea for its operator (see figure 5.4). Supporting users to extend their Microbees were the magazine *Online: The Microbee Owner's Journal*, and—published in 1985—the *Microbee Hacker's Handbook*, the advertisement for which promised "hard and soft projects for Bees of all vintages. For Bee owners who like to . . . put their soldering irons to use" (*Microbee Hacker's Handbook: Hard and Soft Projects for Bees of All Vintages* 1985) (figure 5.5).

The preceding accounts, advertisements, and the *Microbee Hacker's Handbook* itself (see the very evocative cover in figure 5.6, where the user's monitor is encased within a Kambrook kettle box) point to the existence of a strong electronics and engineering ethos in early hobbyist computer culture. The reference to soldering irons, as well as the humorous depiction of hacking contained within the ad—a man taking to a computer's innards with a mallet and screwdriver—highlights once again that users were interacting experimentally with computer hardware, not just writing software. The *Microbee Hacker's Handbook* paints a humorous portrait of the obsessive hacker who was forever "adding things on" to their computer. It read, in part:

> Haven't you always thought that your Microbee could be the very best machine ever . . . if only it had a proportional analogue joystick? . . . (hammer, hammer, hammer) . . . And if it had a parallel printer interface . . . well, it follows that you could hang a parallel printer off the side, doesn't it? . . . (bash, bash) . . . And that printer would really be earning its keep if you could somehow wire up a phase-locked loop decoder, a pitchpipe tuning aid and a shortwave receiver, to the Bee, so you could receive the signals that would let you print out weather maps . . . a bit of Clag should hold it . . . hmm. . . . There must be room for a . . . um . . . ROM reader to plug in the back there somewhere. (Harrison 1985b, 59)

This lighthearted treatment of the will to mod one's machine, for its own sake, captures something of the irrepressible curiosity and desire to be trying new things—once again, the intrinsic motivations that seemed to drive those who tinkered with early microcomputers.

While it is impossible to arrive at the number of microcomputer electronics hobbyists, it is clear that tinkering with the electronics inside one's computer was not limited to people who already possessed electronics skills.

Figure 5.4
Advertisement for the Microbee "Experimenter," with robot tea-pouring arm, *Your Computer*, vol. 3, no. 5, December 1983, p. 105. Courtesy of Ewan Wordsworth, Microbee Technology. Collection of the State Library of NSW.

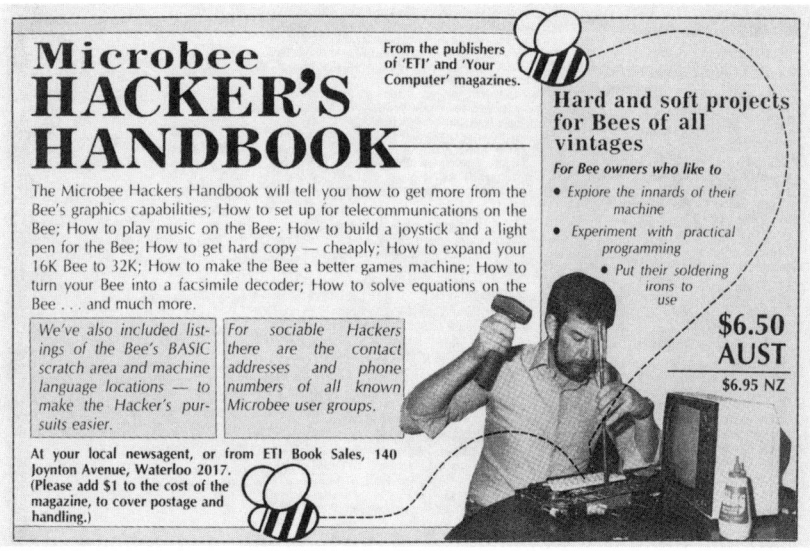

Figure 5.5
Advertisement for the *Microbee Hacker's Handbook* in *Online: The Microbee Owner's Journal*, October 1984, p. 40. Collection of the State Library of NSW.

If a user needed to fix their computer, for instance, there was probably a book for their model of micro that would walk them through the process. Like the encouraging tone of the many teach-yourself-to-program books discussed in chapter 2, users were implored to have a go despite having no prior experience. John Heilborn's book *Commodore 128 Troubleshooting and Repair* is a case in point. The book extends the general invitation and encouragement to tinker (found in electronics and computing magazines) into the realm of repair and troubleshooting. Heilborn[7] claims to cater to "both experienced and inexperienced users who want to repair their Commodore 128" (Heilborn 1988, ix). He writes that he assumes no knowledge apart from "that you are reasonably handy and that you want to fix your computer. I'll discuss the use of any tools you'll need as we come to them" (4). Among other things, the book explains resistors, transistors, integrated circuits, and clocks and has diagrams of the internal architecture of the 8502 chip and locations of the 8502 and Z80 on the processor board. Other books in the series include Heilborn's *Microwave Oven Troubleshooting and Repair, Printer Troubleshooting and Repair,* and *VCR Troubleshooting and Repair Guide*. The encouragement to tinker and troubleshoot that such books

Figure 5.6
Cover of the *Microbee Hacker's Handbook*. (See plate 13.)

provided demonstrates that microcomputer users engaged in a remarkable variety of activities, including picking up skills and knowledge that users are presumed not to have.

Where Is This User?

My central thesis in this book is that the early microcomputer user was a particularly strong example of de Certeau's insight that users and consumers are makers and producers of both culture and artifacts. The electronics productivity of users is further evidence of this. Yet where does this user of the 8-bit microcomputer era—strongly invested in electronics and engineering—appear in cultural, media, or fan studies? The simple answer is that they don't.

Fan and media studies are not alone in having overlooked the micro user as builder and hacker. With only a few exceptions, mentions of early users' electronics nous and hardware hacking in computer history have also been scarce, perhaps because users and consumption have not been considered priorities of the discipline up until now. Computer historians have shown little interest in the history of everyday microcomputing, with personal computers and microcomputers only occasionally featuring as a topic of study. Patricia Galloway is one of the few who mention users' expertise and backgrounds in fields such as ham radio (Galloway 2011, 2011b; Saarikoski and Suominen 2009; Haddon 1988), while Honghong Tinn's work on the production of Apple II compatibles in Taiwan is a notable exception. Though Tinn is largely concerned with how the production of compatibles shaped the social meanings of microcomputers in Taiwan, she prefaces this with a rich account of the amateur electronics hobbyists who assembled their own microcomputers from components chosen from computer shops at a fraction of the cost of a branded computer (Tinn 2011, 75–77).

The neglect of humanities perspectives in engineering is a topic of debate at present, but the opposite also seems to be true, with the engineering connection too far afield for many humanities scholars to venture into (or at times even notice). Therefore, the remarkable electronics ingenuity and hacking and building activities that some micro users engaged in during the 1980s—the cultural practices of repair and improvement and the refashioning of the sociocultural environment, the significance of which de Certeau and Giard urge us to recognize (de Certeau and Giard 1997, 113–114)—have received scant attention from humanities scholars, despite significant interest in user production. As Jenkins writes, "Everyone's talking about consumers as active participants" (Jenkins 2007, 361). Despite this, the technically competent 1980s micro user has been neglected, all but forgotten. It's a curious omission—or blind spot even—within fan and media studies scholarship.

I venture four interimplicated reasons for why this user has been overlooked. First, user creation with micros is, as I have demonstrated, ordinary culture. This includes practices of microcomputer building, hardware hacking, modifying, and troubleshooting. Much scholarship of digital media use and engagement has, by contrast, tended to center on popular culture. Recall that de Certeau and Giard make a distinction between ordinary culture and mass (or popular) culture; ordinary culture is easily overlooked because "our instruments of analysis, modeling, and formalization were constructed for other objects and with other aims" (de Certeau and Giard 1998a, 256).

Second, while the concerns of different scholars working in the field of user production vary—fans' engagement with popular media; the implications of user-generated content for identities and communities; the cultural transformations that are perhaps heralded by user practices; the implications of sharing and participatory culture for learning; and so on—the dominant paradigm of what it means to be actively creating largely remains reading or writing, or quasitextual content production. In general, scholars of digital media consumption have not ventured very far beyond this model.

Steven Jones provides a neat example of these two tendencies in his book *The Meaning of Video Games: Gaming and Textual Strategies*, in which he discusses fan appropriation as part of, or belonging to, a culture of videogames and gamelike expressions of popular culture. He writes:

> The activities of fans as "textual poachers" have been thoroughly explored by Henry Jenkins, who looks mostly at the example of TV shows and draws on Michel de Certeau's notion of cultural poaching to describe the appropriative and reconstructive energies of fan culture. But I would suggest that the source of these fan appropriations, by the 1990s at least, was less a general, overarching "postmodern sensibility" [Celia Pearce's argument, referenced earlier] and more the specific culture of video games and gamelike expressions of popular culture, as these thrive on the Internet. This general ethos is apparent in "Web 2.0" technology developments—podcasts, blogs, Wikipedia, Flickr, Google Maps, Facebook, social tagging, social software applications of all kinds, all with user-created or user-aggregated content. Earlier forms of fan culture, first of course pre-digital and later using message boards, mailing lists, and websites, were precursors (and arguably sources) for these

developments. Steven Johnson refers to the tie-in fan-based websites for TV shows, "online media that latches on to traditional media," using a term that connects suggestively with what I have been saying about paratexts, as "para-sites." Arguably, such community-driven fan media reached a critical mass among gamers, who were among the first in digital culture to appropriate and repurpose the cultural products around which their community was formed, by modding the games themselves, adding custom levels, and producing paratextual media objects such as machinima films made within the gameworlds using the game engines.

But by now, I think, such acts of appropriation and repurposing . . . of a cultural product appear increasingly like the norm in popular media. (Jones 2008, 45)

Jones uses the example of video games as not just a form of popular culture but suggests that popular culture is becoming more videogame-like before characterizing engagement with various forms of popular media as paratexts. The analysis is not wrong, but the point I would make is that not only do such practices have a much longer history than many people know or appreciate but at least some of the practices are not easily wrangled into a reading-writing paradigm. Long before message boards and mailing lists, before tie-in websites for TV shows, and well before 1990s hit games such as *Doom* and *Quake* offered players the chance to mod, vernacular digital cultural production was taking place as micro users developed their own games at home.[8] And as part of this vernacular digitality, some users were producing cultural artifacts that required (and developed) quasiengineering competencies.

Third, there are some significant unacknowledged debts that have shaped thinking in user production studies that have contributed to the lack of attention to the activities of building and hacking. In addition to scholarship coming out of a cultural studies discipline that was not sufficiently cognizant of its debts to literary studies and writing on the one hand (de Certeau's "scriptural economy" of volume 1 of *The Practice of Everyday Life*), media studies partly inherited the cultural studies mantle in the first decade of the twenty-first century but without adequately acknowledging its debt to cinema and film studies, spectatorship, and visuality on the other and the bias toward the screen that results. These have all been useful traditions; however, the weight of their shared inheritance—the allegedly "passive" spectator before a screen—can sometimes seem inescapable. But beyond that, these traditions are not well equipped to notice or theorize users with deeply technical engagements who make cultural artifacts.

Finally, there is the historical moment in which these disciplines were emerging. Two details are important here. The first is that cultural, television, and fan studies were relatively new fields undergoing formation in the 1990s and the decade that followed.[9] It is therefore not surprising that the examples of digital media consumption that have tended to be studied and developed into disciplinary touchstones date from this period. Arguably one of the most influential titles is Jenkins's 1992 monograph *Textual Poachers*, which of course borrowed the figure of poaching from de Certeau and—blended with reader-response theory and ethnographic fieldwork—applied it to fan activity. *Textual Poachers* became a pivotal work in television, spectatorship, and fan studies (Jenkins 1992).[10] The other significant point about the timing of cultural and fan studies' emergence is, of course, the coming of the internet, which was treated—or rather came to be constructed—as a watershed moment with regard to digitality. From the mid-1990s, scholarship tended to cleave around the popularization of the internet and other new forms of digital media, particularly networked media. The coming of the internet might be one watershed moment in vernacular digitality,[11] but there is an earlier one. In chapter 3, I cited Matt Hills's acknowledgment that emphasizing the newness of user-generated content downplays longer histories of user-made websites, and Paul Booth's awareness of pre-internet histories of fan-generated material, but I would argue that this doesn't go far enough. Prior to the internet, people weren't creating using only analog means and weren't only creating content. Users were creating digitally well before Mosaic was released in 1993 or Windows 95 saw the mass uptake of computers. And while some in the 1980s enjoyed the literal act of cutting and pasting, others used early desktop publishing programs for their textual creations (Meggs 2016, 571–573). To return to Giard and de Certeau, not only was homebrew overlooked as ordinary culture because "instruments of analysis, modeling, and formalization were constructed for other objects and with other aims" (de Certeau and Giard 1998a, 256) but also because it was pre-1990s practice. The fields or subfields that were forming during the 1990s were neither concerned with nor equipped to notice what people had been doing with computers in the 1980s. Consequently, the proficient home user of the 1980s microcomputer was occluded from view.

Recovering the Electronics Competency

Recovering the electronics competency—the curiosity, activity, and agency of early users—as another branch in the family tree of user studies interrupts the too smooth continuity that can seem to stretch in some accounts from engagement with film and television screens to engagement with computers (sometimes treated explicitly just as screens).[12] Homebrew contains a number of overlooked threads and histories that are significant not only for historical game studies and histories of technology but that add to our understanding of what it is to tinker and create with technology today. Recovering and restoring early micro users and their practices enables us to push the timeframe of users' digital productivity back much further, providing some redress of the limited attention that's been paid to a longer history of *digital* fan practices. Remembering the perspectives of those users who fiddled with, tinkered with, hacked, and modified their 8-bit microcomputers also provides an expanded historical account of use, introducing a much-needed historicity to accounts of user production in general. Finally, recovering this branch of the family tree of user productivity contextualizes a number of recent and contemporary practices historically, such as overclocking, speedrunning, circuit bending, case modding, repair movements, electronics recycling, "teardowns," and electric vehicle conversion (Simon 2007; Franklin 2009; Whitlock 2017), connecting such activities with historical antecedents that make sense. Doing so promises to energize fan and digital media studies' thinking about contemporary audience productivity, introducing new perspectives and generating new debates, theories, and arguments.

I have been writing of practices dating from the late 1970s and 1980s, some thirty-plus years ago. In chapter 6, I bring the discussion of user practices with 8-bit microcomputers into the present day.

6   The Legacy of 1980s Homebrew

This book has been concerned with practices of coding and hardware hacking from the microcomputer era. This is a period that is rapidly receding: 1986 is as many years in the past as 2050 is in the future (Ferguson 2018). This makes the question of how the microcomputer period will be remembered a reasonably pressing one. In computer history, the case for collecting and preserving software and documentation has been made for some years, but the discourse tends to privilege functionality over cultural memory: future historians will need access to our software, it is argued, and software will also be required for opening archival documents created using that software. Such efforts are undoubtedly important for ensuring the future legibility of historical files and archives, and I have advocated for the collection of, and spearheaded research into the preservation of, digital games and other software and varieties of complex digital artifacts in Australia and New Zealand. But arguments for software preservation and access do not fully capture the cultural significance of historical computing, as a number of scholars have argued. Jussi Parikka, for instance, writes that "there is more to archiving software cultures than focusing on the bits themselves" (Parikka 2012; Lowood 2016; Newman 2012; Swalwell 2017a). In this chapter, I consider the legacy of the 8-bit computing era and of homebrew game development practice specifically, applying critical pressure to the accepted wisdom that forty-year-old computing practices are obsolete. What does it mean to ask what the *legacy* or legacies are of a period, a set of practices and technologies? The *OED* offers several definitions of the noun that relate to "bequeathing something":

> 5b. "A tangible or intangible thing handed down by a predecessor; a long-lasting effect of an event or process."

Meanwhile, as an adjective, legacy designates "something left over from a previous era but still in active existence." This chapter covers the influence of the 8-bit era on contemporary game development and the continuity of programming for micros more generally before considering the complex temporal remediations involved first in demakes and later in the use of vintage games for contemporary political expression. I argue that such apparently anachronistic practice is presenting a new discourse on game history: users are demonstrating not only that the 8-bit era is not obsolete but, by making something new with something old, they are exhibiting the dynamic relation between past and present. I discuss the status of homebrew with respect to collecting mandates and outline some of the remarkable collecting and preservation efforts that are currently under way by Apple II enthusiasts. These are not only identifying, preserving, and making accessible software that was produced in the micro era but also making games heritage available for reuse in current contexts.

Legacy for Contemporary Game Development

The 8-bit era holds clear relevance for contemporary game development, particularly the "independent" game production scene. Parallels are frequently drawn between the current moment and the 8-bit homebrew era, with a number of writers on the digital game industry invoking 1980s antecedents. Anna Anthropy, for instance, sees 1980s homebrew as providing a context for the contemporary retro homebrew and indie scenes (Anthropy 2012). Similarly, Tristan Donovan's chapter on the indie games scene is subtitled "Indie Developers Take Games Back to the Bedroom," a reference to the "bedroom coders" of the 1980s (Donovan 2010). A number of my informants commented on such resemblances. Ross Symons, for instance, started off reflecting on the 1980s and then made the link to current market conditions in the global game industry: "I see it as a very romantic time in game [development] . . . and it was the idea that you could make a game yourself. We did the sound ourselves. One person could do the sound; one person could do the graphics; one person could do the code; the same person. And you could write something and get it out there. In fact, it's come around again, I mean you basically . . . you wrote it; you got it to market directly, and we're kind of back there again." Symons is referring to the rise of casual and indie game developers, the closure of many large AAA studios in the wake of the global financial crisis, and the switch to digital distribution as supporting the small-team model of contemporary game development.

Matthew Hall extends this analysis, naturalizing the development of games by a single person or small teams by comparing it to writing books: "You don't generally get books by a committee." Hall elaborates: "Terry Prachet writes a book, [and] it's one author, one book, and for the longest time [for games too], it was Mathew Smith wrote *Jet Set Willy*. One videogame programmer does everything and releases the game and that's the way I always thought it was going to be. [It] just seem[ed] logical." Hall, who grew up coding in the 1980s, reflected on missing out on the single-person production model the first time around, being just slightly too young:

> Video games completely changed. By the time I realised that—looked up and oh suddenly, you know, there's an artist and a producer and a programmer—it was gone. I sort of just had the talent enough where I might have been able to enter the industry but it was five years too late, because the publishers had stepped in. It was big business. . . . Did you see *From Bedrooms to Billions*, the documentary? Exactly like that. The moment when they all got shut out, and these were the top tier programmers. That was when I was coming in. It's like "well I'm completely screwed, I can't do this" and that was when I sort of essentially gave up on the dream and went to university; just went, you know, "I just need money, I need to get a job." . . . But then in 2008, 2009 essentially it's all come around again and suddenly, you know, one person can make a game again. And from the moment that was possible, I left Tantalus to do that. So that was when I formed KlickTock and made games as one person which is sort of what I always wanted to do.

Perhaps because he'd missed out the first time around, Hall was keenly reading the industry signs and felt that he saw the conditions of possibility for a one-person style of game development reemerging earlier than other observers did. As he explains:

> I saw it before everyone else did, that's for sure. Yeah. I mean the moment I saw a PopCap game . . . that looked like it'd been made by one or two people, and it largely was—I mean *Bejeweled* is not a large exercise. And then Big Fish games was around and a lot of those games were being made by very, very small teams . . . 2008/09 was when *Rag Doll Kung Fu* came out, so that was the first third-party game published on Steam. So it was all sort of happening. I was looking at that a couple of years prior and when the first successes started to appear, I left. And . . . I was trying to convince everyone we should be doing smaller, more interesting games but none of them would have any of that. And so yeah I left and it took a few years before I was able to make a pay packet out of it but [I] eventually made it to that place.

The comments of Symons and Hall—stalwarts of the Australian game-development industry—counter the assumption that legacy (as an adjective) entails discontinuation or supersession. Rather, these men are seeing a return of sorts, something that is perhaps more cyclical than linear. That the industrialization of game production that began in the latter part of the 1980s would give way to smaller operations late in the first decade of the new century is certainly ironic. The so-called indie turn has, of course, not been without its controversies, but one of my hopes for this book has been that excavating the hidden histories of homebrew might provide a prehistory of indie or—perhaps better—"informal" game development. I borrow the latter term from Brendan Keogh, who is extending Ramon Lobato's work on informal economies of consumption and distribution in the film industry. Keogh writes, "What I think we are seeing in the present moment is a re-emergence of informal videogame development as legitimised and validated and visible, especially in Western countries but ultimately in regional contexts" (Keogh 2017; see also 2019). Where once the high production values of commercially produced software seemed to displace the pleasure and appreciation of its homemade equivalent, there is currently a resurgence of people who value and enjoy homebrewed game product. Such developers and their supporters resonate with Keogh's point—and my argument in chapter 3—that we need to get beyond valuing games only in terms of economic return.[1]

The influence of the 8-bit era is certainly evident in some contemporary games, whether via an experimental aesthetic or elements such as game mechanics.[2] In 2012, game developer Jim McGinley spoke of the potential for developers to gain "inspiration from the Trash," reviving mechanics and other ideas from games made for the TRS-80, a micro that was affectionately known as the "Trash 80" (McGinley 2012). Hall is part of Hipster Whale, the company behind the highly successful casual game *Crossy Road* (2014) as well as a licensed version of *Pac-Man*, *Pac-Man 256* (2015), a game that he says was "made for retrogamers." He is well aware of the nostalgic appeal of revisiting earlier games, observing that "people have always been interested in the things that fascinated them as kids." Retro inspiration is readily apparent in *Crossy Road*, which is a homage to *Frogger* (1981), though Hall also cites *Flappy Bird* (2013) as an important influence. In 2015, he offered, "There are 100 million players. Probably not even . . . a small fraction of them played the original *Frogger* or even knew what that was. But yeah it's not completely faithful: it's re-adapting those things and making them modern, and, you know, making *Frogger* funny, which it never was." *Crossy Road* is a free-to-play game that has been ridiculously successful, with more than 200 million downloads.

Hipster Whale's games demonstrate McGinley's thesis that the 8-bit era holds considerable inspiration for contemporary developers to mine. The *Crossy Road* example yields more still (in the next section, I discuss Bob Smith's demake, *CroZXy Road*, as an example of contemporary homebrew game development for 8-bit computers). While it is true that in the example cited the designers derive influence from the era broadly rather than from specific homebrew titles, it is inconceivable that adaptations of legacy homebrew have not been undertaken. Indeed, that some commercial game developers began as homebrew developers in the 1980s makes it likely that they would have revisited some of their early ideas over the years.

Anachronistic but Not Obsolete

Apart from the direct legacy of 8-bit game mechanics and games of the era providing inspiration for contemporary game development, 8-bit *platforms* continue to see activity and engender a range of practices now, some old and some new. It is not uncommon to hear people say things such as "the spirit of 1980s home coding lives on in the Raspberry Pi today," and yet the practice of those who continue to code for vintage computers has not received much attention. Part of the reason for this is no doubt because it seems counterintuitive to speak of such practice in the present tense. Yet, despite the naysayers, some users still tinker and code with 8-bit computers, deploying their deep knowledge of the coding routines and systems they knew and loved in the 1980s. They variously produce new games for 8-bit platforms, "demakes," "deprotect" software titles so they can be preserved, and build emulation solutions. Some will find it hard to fathom such anachronistic practices, taking the "dead media" label at face value and believing that micros really are obsolete (Sterling n.d.). Or perhaps they struggle to associate tradition with something as apparently "new" as computing. Notwithstanding such contradictions, in this section I hone in on the temporality and continuity of practices associated with 1980s programming and homebrew game development.

Twitter Straw Poll

I took the opportunity of piggybacking on a story that Attila Egri-Nagy shared on Twitter—with the comment "I'm still coding, just to stay close to the machine"—to reach out and ask other people who still code for retro computers why they do it. The question elicited some fascinating responses, once people realized that I was genuinely interested in their perspectives and not seeking to pathologize them or their practice (Swalwell 2017b). Motivations seemed to fall into five main categories, as follows:

- For fun and enjoyment: "just for fun" / enjoyment / as a break from everyday life / fixed platform for a shared experience; "reverse engineering programs"; "the BASIC prototyping cycle [is] shorter than in modern languages [so] it's fun to tinker, make changes, and see them immediately" (cf. Beals); and because some tasks are more efficiently achieved in an early programming language.
- The challenging constraints of early microcomputers: "the challenge of making something great in a very limited environment"; "limitations spur creativity"; art; "comfortable familiar as a reprieve"; "doing things barely possible, and challenging myself to distil ideas to their essence"; "coding in its purest form, no abstractions."
- The social aspects of coding: Programming contests received a mention, as did "preparing for vintage computer festival exhibits." Interestingly, though vintage festivals and get-togethers might be one of the most visible moments when these are on display, most responses did not mention such extrinsic factors.
- Relatedly, the simplicity of micros: "to get my head around working in assembly language," which was much better done on an early machine than on a contemporary one; "the old computers were fun because they were simple enough you could literally understand *everything*."
- And finally, the fact that time has passed: "because now I know a lot of things I didn't when I had the computers in the 80s, so I am doing things that I used to think [were] impossible"; to make something for a much-coveted machine that has only been recently acquired; "long stewing ideas"; because decades later, there are still things to learn.

These users refuse to "time capsule" (Guins 2014, 3) 8-bit machines and are looking forward, not back. Rather than being characterized by discourses of loss and lament in line with what I identified as decline theses in

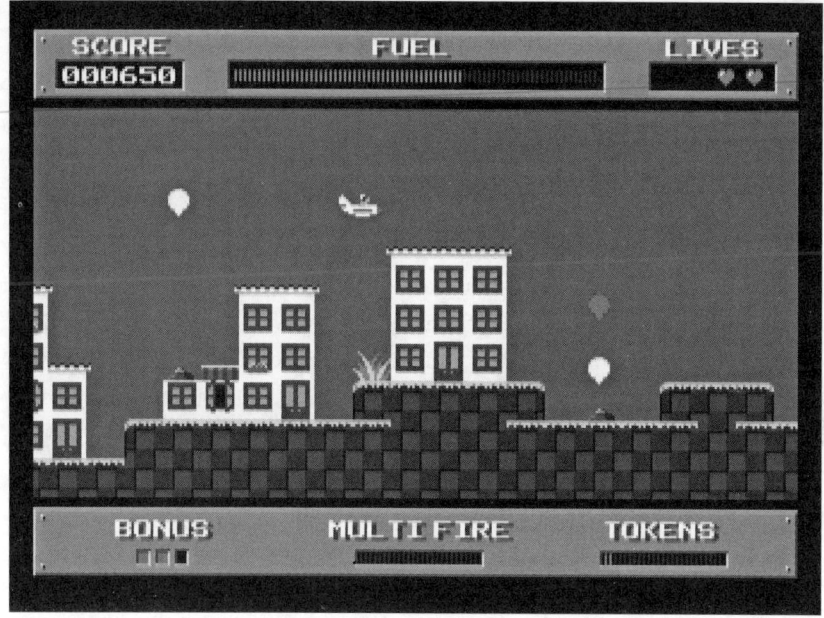

Figure 6.1
*Pop*Star Pilot* screenshot. Courtesy of Nickolas Marentes. (See plate 14.)

chapter 4 or being in the grip of a nostalgia that "sinks . . . efforts to create things that feel new" (Hilbert 2004, 57), I will argue that they are bringing the past into a dynamic relation with the present (Seremetakis 1996, 4) and indeed the future. In what follows, I drill down into interviews with two current homebrew developers making games for vintage platforms, Nickolas Marentes and Bob Smith. Marentes—whose period games I discussed in chapter 4—continues to code for his beloved Tandy Color Computer 3. He describes his practice succinctly: "in essence I'm doing something new on something old." To begin, I discuss his newest game, *Pop*Star Pilot* (2016), followed by an earlier 3D game, *Gate Crasher* (2000).

*Pop*Star Pilot*

When I interviewed Marentes in 2013, he was working on a horizontal side-scrolling game that he had called *Pop*Star Pilot* (figure 6.1), though he estimated it was at least a year from completion. He explained his motivation:

> Over the years you always pick up other ideas . . . that you've thought up, or games you've seen that you thought, "Gee, I wonder if my computer could do that?"

especially games which are, say, on more powerful computers or computers that came out later which have certain abilities that your computer doesn't have. You see all these things and think, "Well, I wonder how I could do what they're doing on my computer and show them that my computer is not as lame as what they think it is?" So you get all these ideas together, but of course you never have time to go ahead and prove it or to code something together. . . . So basically I'm now looking at these ideas that I've had in the past in this current game from a technical point of view.

Back in the day, he recalls, "I could pump out a game in, say, three months—I'd have the game start to end—whereas now [it takes] more like a year, at least." Having had the benefit of thirty years to reflect, but with less discretionary time than he had as a teen and young adult, Marentes decided to take all the new techniques he's come up with and put "as much as I could into the one game."

*Pop*Star Pilot* embodies his desire to make a game with really smooth split-screen side scrolling, "a feature that this computer could do, but not quite. It does it [but] with limitations." He explains:

It's been rattling [around] in my head, okay, and I've worked it out and I just have to do it. So basically I've got that idea, then another idea with sound effects; how can I create sound effects, great digital sound effects that doesn't interfere with the animation? You come up with all these ideas . . . and I've pieced them all together. . . . And that's what this game does. It utilises smooth split screen hardware scrolling which is not a feature of this computer. It uses digital sound effects simultaneously with the animation, something that this computer is not meant to be able to do without any additional support hardware, and there's a few other little aspects in there I also wanted to explore.

At the time of writing, Marentes has sold over 120 copies of the game via mail order. His comment is characteristically dry: "Pretty good for a 30 year old dead computer. :)" Sales figures aside, the contemporary retro homebrew scene is where we see the potential of the 1980s experimental ethic fulfilled most clearly: that is, practice that is curiosity driven, with users interested in seeing what is possible and gaining an appreciation of new programming tricks, as well as interrogating hardware and pushing it to its limits—what I describe to Marentes as a "demoscene ethic." Such an ethic is even more evident in his earlier game *Gate Crasher* (2000).

Gate Crasher

After his *Pac-Man Tribute* (1997), Marentes took a couple of years' break before returning to programming his beloved Tandy CoCo. In 2000, he released *Gate Crasher*, a 3D game for the 512K Color Computer 3 (figure 6.2). *Gate Crasher* is a demake, because Marentes is taking the concept of a 3D

Figure 6.2
Gate Crasher screenshot. Courtesy of Nickolas Marentes. (See plate 15.)

first-person shooter game back to a platform that was only thought to be capable of running 2D games. (The term *demake* has been adapted from *remake*, but rather than indicating a revision of a title for new hardware, it connotes a step back in terms of the computer hardware that is being developed for.) Back in 2000, Marentes recalls, people were saying that there was no way you could get a proper 3D game for the 2 MHz computer: "Even in machine language it wasn't fast enough to do all the mathematics required to recreate the 3D environment, especially one that you can freely move around in. There were 3D games, but they had a very fixed 3D frame. You'd look down a corridor and it showed you the perspective down that corridor, but if you turned left you would suddenly just turn a full 90 degrees and look at the perspective of the next corridor, say. You couldn't just move a little bit to the left or walk slightly to the corner of the same room."

As Marentes tells it, Canadian John Kowalski changed all that with *Gloom* (1996), a graphics demonstration engine, saying that this was "a demo that actually showed a 3D environment that you could freely walk around in" (Kowalski 1996). Despite Kowalski's impressive feat, Marentes recalls, "people were saying, 'Well, that's very good but you couldn't use that in a game because in a game you've got to include sound effects, you've got to include scoring, you've got to include gameplay. . . . And as soon as you add those in, your frame rate will drop and, well, there goes your 3D effect. It'll be so slow it's just pointless to actually have a game on that hardware.'"

Marentes took this as a challenge. He befriended Kowalski, who explained his algorithm. Marentes then went and used this on the demake *Gate Crasher*. Some sacrifices had to be made: as the doubters had foreseen, Marentes had to add in a range of game elements—characters, a story line, scoring, and sound effects—so he used a lower graphical resolution and took a few other shortcuts that mean that his game doesn't run as smoothly or as fast as Kowalski's demo. But the result is surprisingly good and incorporates most of the basic elements of the first-person shooter genre, on the Tandy CoCo 3, a computer that first appeared in 1986, well before such classic 3D games as *Wolfenstein 3D* (1992) or *Doom* (1993).

While Marentes is completely committed to maximizing the performance of the Tandy computers, his approach to coding has mellowed somewhat over the years. It is now very much a labor of love—"a long-term hobby type project"—which he does primarily for the challenge. A game takes him much longer to write these days, but, being an early riser, he does all his programming in an hour or two early in the morning while "everyone else is still snoring."

CroZXy Road

> Taking it to the ZX81—you're going the wrong way!
> —Philip Oliver of Smith's *Ant Attack*

Bob Smith also busies himself with the art of the demake.[3] In the late 1980s, when he was fifteen or so, the British teenager began typing in games from magazines on a Sinclair ZX81 and later programming a Spectrum in BASIC. Before long, he started playing around with the programs and seeing what he could write himself. After he'd learned BASIC, he started on machine code, but by the time he'd started trying to write his own games, the Spectrum was at the end of its life (or so he thought). He went off to study computer science at university, followed by some twenty-odd years working as a programmer in the game industry. But it seems the 8-bit "itch" never left him.

Decades later, Smith got down from his parents' attic the box containing his Spectrum and the game he'd written. *Stranded* was published on tape in

2005 by UK publisher Cronosoft. Smith wrote a few more titles, but at that point, he recalls:

> The Spectrum didn't have much of a following: there wasn't much being produced for it, it was quite quiet. . . . I thought, why not, why not go back to where I started? So I started looking at the games you could get [for the ZX81] and found they were all really awful. They were modern written games but they were all very much in the style that you would have put out in 1980, very unresponsive, very slow . . . just sort of using letters to build shapes and this sort of stuff. . . . I thought, it's got to be better than that and that's why I started trying to see if the ZX81 could do good stuff and hopefully [I've] proved that it can.

Smith develops for the ZX81 because he is convinced that it should be "capable of doing something better than it was 35 years ago." He explains:

> Many people's memories of it is of this horrible black box that had no keyboard and the games were awful and everything was awful about it and I kind of wanted to address that and say that it was a very important machine. And a lot of people did start their careers on a ZX81 or a ZX Spectrum.
>
> But I think it was passed by and forgotten about since the Spectrum got launched . . . which is fair enough because the Spectrum is a far better machine.

As Smith writes on his website, "Despite being a professional programmer since 1994, and employed in the games industry since 1996, I still need another outlet for my games ideas and programming talent" (Smith n.d.d). When Smith is asked about his motivations, he agrees that as well as making better software for the ZX81, he wants to try and see what he can get out of the hardware: "It's a . . . 'geek sudoku' sort of thing because it's such a limited hardware, 16k now is nothing, especially when you have to . . . [when you] think that all the game and graphics all have to live in 16k. It's quite difficult to begin with, even to get a game into 16k and then write it in such a way that [it will run] on a—by modern standards—sort of prehistorically slow machine . . . as a fluent, responsive game. That's a great challenge."

Getting the very limited hardware to do things that were never thought possible through clever programming tricks is not the only way in which Smith's practice resembles Marentes's. Both have written their own versions of "classics" (Swalwell 2016). "*Impact!* for instance," Smith writes, is "my interpretation of Atari's 1979 arcade game *Asteroids*." And where Smith adapts an idea, like Marentes, he seeks to improve on it, as in the case of *Noir Shapes*, of which he writes: "Following from the excellent *Miner Man* another conversion of one of Electric Wolf's Xbox 360 games to the ZX81. Again, it doesn't have some of the in-game features, any sound, colour graphics, and such like, but this time we have managed to improve upon the original by adding an extra 12 levels to take the total to 60, and have even converted the use of an avatar from the original!" (Smith n.d.c).

Neither Marentes nor Smith are purists as far as 1980s development practices go. While Smith loves the challenge of developing for old micro hardware, he is not worried about the authenticity of practice and uses some contemporary development tools. He explains, "When I did my Spectrum game back in the day it was all little chunks of code and I'd written it all out on paper and graph paper and things and typing it all in by hand. In the modern way of doing things, I've got an assembler and a development environment and I can use other packages to make things, so it's a lot easier now."

Smith also uses a PC, saying, "There's no way I think this would ever happen on a normal ZX81, because even loading stuff can take [time]. It'd be hopeless." Some of his more recent titles for the ZX81 utilize the Chroma interface, a peripheral that allows the monochrome micro to be connected to a TV via a SCART socket, producing "an RGB picture that is sharp and bright" (Farrow 2014). A complete anachronism, as Smith says, this means that in *U-Bend* "the water is actually blue!"

In 2015, Smith's attention was captured by Hipster Whale's *Crossy Road*. For Smith, it was "arguably the most iconic game of 2015—instantly recognisable, simple to play, and ruthlessly addictive—and so it seemed only fitting to try and bring it back down to Earth, and back in time by 35 years, in the form of a de-make for the ZX81" (Smith n.d.b). Hall cut his teeth coding on a Vic-20, so a demake of the game for an 8-bit platform has a certain rightness to it. *CroZXy Road* offers a complex remediation of genres and computer hardware in a bizarre twist on McGinley's thesis that contemporary developers find inspiration in 8-bit classics. And while *Crossy Road* is itself quite a funny game—referencing the pointlessly absurd humor of road-crossing chickens—the thought of the brightly colored, simple, free-to-play casual mobile game "regressing" to make an appearance on the much older ZX81 is hilarious (figure 6.3). Smith acknowledged that "it's somewhere between genius and lunacy [on his part], even attempting [the demake]."

Not only have developers been happy to support Smith's efforts, but his habit of seeking their blessing for his adaptations has led—in the case of his *Ant Attack*—him to use some of the original code base from Sandy White's 1983 game.[4] Smith writes that when he was playing around with doing a 3D isometric game,

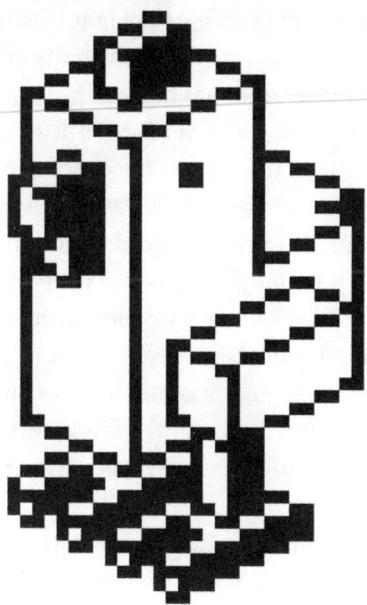

Figure 6.3
CroZXy Road chicken. Courtesy of Bob Smith.

> *Ant Attack* quickly became an obvious choice . . . but if I was to do it justice I had to somehow get a 48K Spectrum game into the humble 16K of the ZX81, and I knew that the map data alone consumed 16K of the original. Initially, I wrote a map viewer to show a compressed version of the map on the ZX81 as a proof-of-concept, and posted a picture of it into the guest book on Sandy White's website, along with a brief explanation of what I was plan[n]ing to do, and asked if he was OK with the idea. . . .
>
> Luckily he was, and more than that we started to work together on the idea of using some of the code from the original in order that this not just be a remake but as close to the original as possible. Over six months later, and numerous redesigns of my code as I repeatedly ran out of memory, and the game is finished. I'm really pleased with the result, not just from a coding stance but also in my appreciation of the original game, how it was coded, and just what an amazing game it still is. (Smith n.d.a)

Smith writes that his *Ant Attack* ended up being "a sort of . . . proper official copy, as it were."

Hipster Whale was also happy to support Smith's *CroZXy Road* project, provided he wasn't intending to commercialize it. They even helped to name the game. I get the sense that Smith pushes himself very hard to ensure that his version of a game is a worthy homage. I ask him whether the term *clone* is apt. Smith prefers the term *demake* to *clone* to describe his version of the game, explaining: "I think *Crossy Road* has been cloned enough on the Android and App Stores . . . and there's 101 versions of it that don't play like it or just [try to] to piggyback on its success." By contrast, Smith wants to prove that it's a very good game—"from a game design point of view, it's very solid"—and that it can work on other platforms. He elaborates: "It's very easy to understand, it's very simple to play and yet very difficult to be any good at it. I'm hopeless at it, but everybody can play it and I think that's something really special about a lot of these [casual] games. . . . You don't have to learn complicated control systems or lots of different button presses or anything, it's very simple to play. So I think it's just a great game and it's got this dark humour about you getting splattered on cars and things so it's just a lovely game."

Smith would like to think that Hipster Whale is happy with *CroZXy Road* and also impressed at the achievement. When we spoke, he had the sense that Hall was impressed because he remembers the period and the realities of programming within limited memory. There's a circle of admiration that seems fitting here: Smith the programmer admires Hipster Whale's game design, and in an act of homage, creates a demake, which is in turn admired for its virtuosic technical achievement.

In 2007, I published an article, "The Remembering and the Forgetting of Early Digital Games: From Novelty to Detritus and Back Again," in which I argued that there was an urgent need for discourses reflecting on digital games in relation to broader shifts in visual culture. It essayed what I characterized as the "abundant contradictions between games' early novelty, their subsequent rejection, and a more recent (partial) recuperation of these artefacts—a cultural position that is thoroughly ambivalent, incorporating excitement, nostalgia and amnesia" (Swalwell 2007, 256). I invoked Tom Gunning's idea that for technological novelty to be noticed requires a discourse in which it can be expressed (Gunning 2003). Gunning outlines the importance of discourse not only in shaping reception—as his work on the alleged reactions of early audiences of the Lumière Brothers' films shows (Gunning 1989)—but in novelty being noticed in the first place. Citing Victor Shklovsky's futile search for accounts of the introduction of electric light to Moscow and Petersburg, Gunning concludes that "journalists lacked a discursive context, or tradition, for the expressing of such astonishment"

(Gunning 2003, 44). Applying Gunning's arguments to the case of early digital games, I pointed to the lack of a discourse, beyond nostalgia, for games' recuperation. In 2007, when I was writing, nostalgia had been the dominant motif of remembrance for more than a decade, and at the time it seemed to me that nostalgia was the only extant discourse. Now nostalgia often gets bad press among game historians. Oftentimes this is attributable to fannish apprehension of game objects. However, the disdain in which nostalgia is sometimes held also stems from a narrow understanding. By contrast, I invoked Nadia Seremetakis's definition of nostalgia, where she excavates the Greek roots of the term that tie nostalgia to sensory memory: "'*Nostalghía* speaks to the sensory reception of history.' Yet Nostalgia, in the American sense, freezes the past in such a manner as to preclude it from any capacity for social transformation in the present, preventing the present from establishing a dynamic perceptual relationship to its history" (Seremetakis 1996, 4).

I want to suggest that micro users such as Smith and Marentes—who continue to deploy their deep knowledge of, and *love* for, the coding routines and systems they learned in the 1980s—are now, through their actions, effectively articulating a new discourse around game history. They are not themselves in the grip of a nostalgia that, to borrow Ernest Hilbert's phrase, "sinks . . . efforts to create things that feel new" (Hilbert 2004, 57), nor are they encouraging this in others. Rather, they are—quite literally—doing what Seremetakis gestured toward: bringing the present into a dynamic relation with the past.

Developing games in 2016 for the Tandy CoCo or the ZX81 may seem anachronistic, but Marentes's and Smith's practice demonstrates the very lively legacy of 8-bit micros and games now. Theirs is a different kind of technocultural legacy: rather than just revisiting the past and attempting to somehow keep it *how it was*, what these creators have made feels new. They lived the era, programming the first time around, and have now had over thirty years to reflect on what else can be done. Through their innovation and virtuosity, they are making historic computers perform new feats, in a sense reinventing the platforms for which they code, and they're doing it with contemporary eyes. To riff on Benjamin, "brushing history [so] against the grain" potentially makes this period of game and microcomputer history newly accessible, giving audiences "a unique experience with the past" (Benjamin 1992, 248, 254). *CroZXy Road* potentially speaks powerfully of

8-bit computing—of a time when it was possible to teach oneself to code and write a game during the school holidays—to a generation that knows *Crossy Road* but never played a ZX81.

To wit, my daughter accompanied me to the Berlin Computerspiele Museum when she was eight, playing an arcade version of *Frogger* during the visit. She is growing up in the era of computer ubiquity, when school rolls are marked digitally, our digital personal data are matched with abandon, and there's an app for almost every conceivable function, but also when the inventiveness of applications—perhaps still motivated by a programmer's avocation, though with a dose of platform capitalism thrown in—is much more blackboxed than in the 1980s. Having played *Crossy Road*, she could completely appreciate the relation and also the differences between it and *Frogger*. Whether she'll get the chance to really get among the "digital sinews," beneath the "slick," "pristine," "polished slabs of glass and metal, performing veritable feats of magic" (Arbesman 2015) and get close to the machine as kids of the 1980s did still remains to be seen. Maybe I'll sit her down with *CroZXy Road* during the next school holidays, alongside her Raspberry Pi, and see what she makes of it. Bringing the present into a dynamic relationship with the past can also work the other way around: retro YouTuber Lord Villordsutch made a Let's Play of *CroZXy Road* and despite not knowing much about contemporary casual games—but clearly quite a lot about the 8-bit era—he was deeply impressed by Smith's achievement ("Bob's worked sodding wonders!") (Villordsutch 2016).

Rather than asking why programmers would make games for vintage computers, it is perhaps more revealing to consider what makes such a practice difficult to fathom. If Marentes's and Smith's creations trouble our categories, then our categories are surely too fixed. History is not static, nor are user practices with the 8-bit computers on which homebrew games were and are being developed. I suspect part of the challenge is also because creations such as *Gate Crasher* and *CroZXy Road* challenge assumptions about obsolescence and the certainty that old ways are in decline, assumptions that cloak a narrative that the new replaces the old and is automatically better. Older practices and technologies are not simply displaced by newer ones, a subject usefully addressed by Mark Thomson in his 2002 book *Rare Trades: Making Things by Hand in the Digital Age* (Thomson 2002). Thomson's thesis is that rare trades such as those of the cooper, the wheelwright, and the stonemason "are not disappearing . . . but persistent trades" (Thomson

2002, frontmatter). When a ceiling rose in a historic home needs restoration, for instance, the artisanal skills of a decorative plasterer will be sought. Such dedicated specialist practitioners are in high demand. Silicone molds might have replaced gelatin, and newer techniques such as laser cutting will be deployed alongside older ones, but the trade is not defunct (152). Transformations in practice have a much longer tail than is commonly acknowledged, with the old and the new coexisting for a good while (see, e.g., Švelch 2017).

Collections

Another factor materially affecting the homebrew legacy and how long it will endure is the response of collecting institutions. It is now less rare for cultural institutions to have some digital games in their collections than when I began working in game history a few years into the new millennium. At that time, there were only a few game archives. I recall being asked to gloss these for readers' information and enlightenment in an article reporting on a game preservation pilot project I had undertaken with colleagues in New Zealand (Swalwell 2009). Happily, that list is woefully outdated, now mostly of historical interest, as more institutions have embraced digital games as an important media and cultural form.

As games find homes within cultural institutions, their legitimacy changes. Yet "ordinary culture" is still easily overlooked in favor of "official culture." As I have argued, a number of factors have historically militated against homebrew being thought significant enough to warrant inclusion in archives. These include homebrew's location in private domestic space, the clone allegation, and the fact that developers mostly made software for low-end computers. While not everything deserves to be preserved, the decisions that are made today about what to keep from the early microcomputing period will determine the narratives and histories that are able to be told in the future. If these are to include stories about what ordinary people did with computers when they first became available, then collection policies and accessioning will need to bring everyday homemade products and producers such as homebrewers into scope. If collecting institutions do not go out and actively solicit such materials, then it is unlikely that they will come to them.[5]

Ironically, I have found that some homebrew authors have better archives than commercial developers do, or at least better than the latter are prepared to share. Elsewhere, I've argued that game development materials such as artwork on graph paper—which creators often used to design graphics (see figure 4.6)—very effectively convey what it was like to make a game in this period (Swalwell 2017a). A form rejection letter like the one Harvey Kong Tin received after Antic evaluated his game *Hot Copter* (later published in 1986 as *Laser Hawk*), developed with Andrew Bradfield (figure 6.4), is surely the counterpart—and corrective—to the celebratory magazine calls for submissions I discussed in chapter 4 promising that anyone can make their fortune writing software. These very personal artifacts convey something of the disappointment that homebrew authors must have felt (Swalwell 2017a; Kong Tin 1986). I have been fortunate to have partial access to the archives of homebrew developers Kong Tin and Marentes. Scans of some of Marentes's archive form part of the Play It Again research collection at the Australian Centre for the Moving Image, while Kong Tin's are online at the Internet Archive. Scattered artifacts and evidence of homebrew activity exist in some archival collections, but traces of vernacular game development are outnumbered by those of official game development and culture.

Unofficial Archivists

Further evidence of the lively legacies of 8-bit microcomputing and games from the 1980s is seen in both the activities of game preservationists and in the activities that game preservation enables. Enthusiasts have long played a central role in game history and preservation. They have curated and preserved games and game history sources through decades of collective unremunerated work (Stebbins's "serious leisure"), creating online archives, building emulators, and forensically analyzing code (Swalwell, Stuckey, and Ndalianis 2017). Disk and tape imaging tools such as the Kryoflux, the 1541 Ultimate II, the Applesauce, and others exist because groups and individuals developed solutions to the challenges of preserving and accessing legacy software by using their in-depth knowledge of specific platforms. As Frank Cifaldi tweeted, "Emulator authors should have statues erected in public spaces. They're the heroes who put in the exhaustive free labor documenting how these old games worked" (Cifaldi 2018). It is increasingly recognized that the emulators built by computer enthusiasts—often to play vintage games—offer an effective means of accessing other historic coded works, including art (Ippolito and Rinehart 2014). Methods and insights from

```
ANTIC®
The Atari Resource
II COMPUTING®
For Apple II Users
ANTIC® Publishing, Inc.
524 Second Street
San Francisco, Ca 94107
(415) 957-0886
```

Dear Software Developer

Thank you for sending us ___HOT COPTER___.
We regret that it does not fit our current needs, so we are unable to add it to The Catalog.

Unfortunately it is impossible to market all the good programs which we receive. Also, the large number of submissions prevents us from replying with specific comments or suggestions on each program.

If you wish to have us return your program and any other materials which you may have submitted, send us self-addressed, postage-paid packaging. We will hold your materials for 45 days, and, if we have not heard from you by then, we will destroy them. No copies will be retained by us. (If you included return postage with your submission, your program should accompany this letter.)

Good luck with marketing your program. When you have your next product ready, please keep Antic and The Catalog in mind.

Cordially,

Charles Cherry
Product Manager
The Catalog

Figure 6.4

Form rejection letter from Antic after reviewing *Hot Copter*. Courtesy of Harvey Kong Tin.

game preservation are increasingly finding a place in media arts conservation (Rieger et al. 2015; Rosenthal 2015; Rechert, Falcao, and Ensom 2016).

Enthusiasts' efforts to collect, image, and emulate historic software have been broadening beyond games for some years now. One significant project that has unearthed some remarkable software artifacts and generated contemporary instances of reuse is focused on software for the Apple II microcomputer. The prize that Apple II preservationists have their eye on is nothing less than preserving all software written for this microcomputer. I want to unpack this project in some depth not only because it highlights the benefits of a participatory collection and preservation project but also because it elegantly demonstrates some of the ways in which the public is invested in digital cultural heritage.

Promoted by the Internet Archive's software curator, Jason Scott, the Apple II project has encouraged people to send in software in their possession for imaging, hosting, and emulation on the Internet Archive. Previous efforts spearheaded by Scott have led to the setup of several online software repositories, including the Internet Arcade and the Console Living Room. These offer the emulation of software in the web browser via a series of ports of existing emulator packages, including MAME (Multiple Arcade Machine Emulator), MESS (an emulator for many console and computer systems), and DosBox (an emulator for IBM PC compatibles running Microsoft's DOS operating system). Emulating software in the browser has lowered the bar in terms of the skills and knowledge users require, effectively "mainstreaming" emulation (Swalwell 2017a). The Apple II project has led to many previously unknown software titles surfacing. The collection currently holds more than 28,500 items ("The Software Library: Apple Computer" 2019), while that assembled by the cracker "4am" currently numbers 1,994 titles (4am 2019), including a remarkable selection of educational software. In addition to the spelling, math, reading, chemistry, and geography titles one might expect, there are foreign language tutors, electronics tutors, at least three titles on homonyms, weather forecasting, aids for people with disabilities, and something called *Bible Baseball*. The enormous diversity of titles being amassed and aggregation of the software metadata begins to provide a partial answer to the question I posed in chapter 2 regarding what users did with their microcomputers in the 1980s.

Apart from assembling an impressive collection of software titles, the Apple II project is also a significant site of innovation, with contributors developing a range of new preservation techniques in recent years. Crackers are working to "deprotect" software, while others are developing new hardware to make Apple II software preservation better. The Applesauce floppy drive controller is a hardware device developed by John Keoni Morris that delivers flux level imaging of floppy disks; crackers have incorporated the Applesauce into their methods and workflow (4am 2018).

Circumventing, or "cracking," copy protection brings up vexing legal issues, and this has often stopped those in more traditional cultural institutions from going down this path. However, the time when rights owners pursued those infringing rights on software for 8-bit microcomputers has largely passed.[6] Most legacy software is no longer commercially profitable, so its significance is largely historical. 4am's verification and copy program for 5.25 inch Apple II floppy disks, called *Passport*, automates cracking by "target[ing] common protection schemes that were reused by multiple companies" (4am 2017a). Scott describes these as "silent" cracks, having "no added screens or credits," adding that "many were cracked before, but with modifications, reductions and crack screens" (Scott 2019). 4am periodically tweets about previously unpreserved titles that have now been cracked and preserved, with such announcements adding momentum and attracting further submissions to the project.

The Apple II project could have been a US-centric effort, but it is not, either at the level of content or personnel. The Internet Archive's willingness to host collections from anywhere is no doubt a factor here; for example, Australian Jeremy Barr-Hyde has contributed ninety-two imaged floppies, and his collection includes many locally written educational titles (Barr-Hyde n.d.). The project's internationalism is further underlined by the development of thematic collections, such as the Apple II X project—"a project for adults where we compile the x-rated s/w [software] for the Apple II and IIgs"—in the words of its coordinator, Antoine Vignau in France (Vignau 2017). Collecting the earliest computer porn for Apple micros ensures the 1980s' chapter in the history of *ars erotica* will not be forgotten. At the time of writing, 127 disk images have been assembled with dates ranging from 1980 to 1994. Judging from the file names, this is an international and multilingual collection, with titles indicating French, German, and Japanese as well as English-language origins (Vignau n.d.).

Patches and Fixes

While the Apple II project and its satellite projects are very exciting and will no doubt support fascinating historical research in the future, one of the most significant aspects for me is that the project is going beyond software history per se. It is also facilitating contemporary creative production and political expression, demonstrating that historic software titles are meaningful resources for users to—paraphrasing Marentes—make and do new things with. Only weeks after shocking scenes of police violence against people protecting ballot boxes in the 2017 Catalan independence referendum, a2_poet patched 4am's crack of *Summer Games II*, inviting players to "play an 8 bit sports game of the eighties celebrating Catalonia Declaration of Independence on 27 October 2017" (figure 6.5 shows a screenshot) (a2_poet 2017).[7] a2_poet identifies as Catalan on Twitter and made it clear that they were patching 4am's cracked version of the game. 4am tweeted, "People are taking cracked 8-bit games from 1985 and patching the binaries for political expression in 2017 AND I AM SO HERE FOR IT" (4am 2017b). Responses suggest that opinions were divided. Vignau seemed unimpressed: "If they work from a copy, I don't care. From an historical perspective, that is horrible: denial of History, of good/bad past events." The story is, however, much richer than Vignau's take suggests. The hack is actually thirty

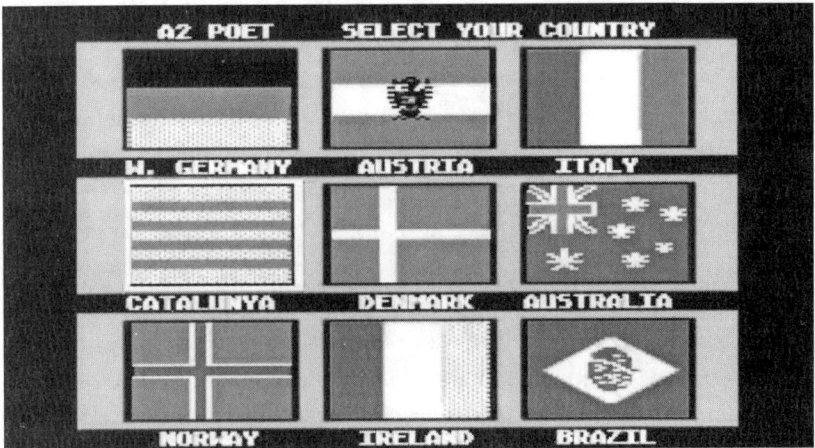

Figure 6.5
Summer Games II (Catalonia patch to 4am crack) country selection screen (2017) by a2_poet. Courtesy of a2_poet. (See plate 16.)

years old, originally performed on the Hot Rod crack of *Summer Games II* in 1987 or earlier. As a2_poet writes:

> The patch consists in being able to compete for Catalonia.
>
> The game allows each player to select one of 17 country flags from the eighties or the Epyx team flag to compete.
>
> As a bonus when the country is selected the national anthem for the chosen country plays.
>
> At the end of each event, the flag of the winner is shown and the national anthem can be heard again through the apple speaker.
>
> To include Catalonia as an option I had to sacrifice another country and guess what . . . the game designers included the flag of Franco's dictatorship instead of the valid spanish flag since 1978.
>
> So the hacker decision was a no-brainer, the catalan flag shares horizontal stripes and colors with the spanish flag: two red stripes for Spain, four red stripes for Catalonia.
>
> The trickiest part was to guess which bytes held the anthems.
>
> Back then I did not know how to write CATALONIA in english so when choosing the flag it read CATALUNYA written

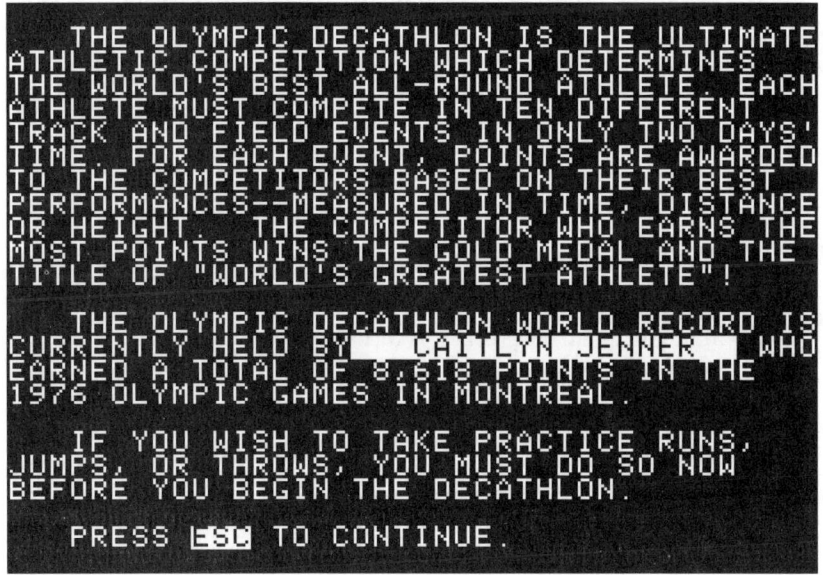

Figure 6.6
Screenshot from "Olympic Decathlon fix" (2015) by Sarah W.

in the Catalan language. This is what this patch fixes.
The country abbreviation (CAT) stood the test of time.

Also, instead of releasing it as a patch to the HOT ROD crack, I was curious if it would work on [the] 4am crack . . . and it worked! (a2_poet 2019)

Broad dissemination of this patched, thirty-year-old statement of self-determination is obviously far easier than it was in 1987.

Also responding to 4am's tweet regarding the Catalan patch, Quinn Dunki commented, "I liked it better when Sarah did it at KFest 2016." The reference here is to "Olympic Decathlon fix" (2015), an entry by Sarah W. in HackFest, an Apple II programming competition held in conjunction with KansasFest, an annual Apple convention held in Kansas City, Missouri. The HackFest archive shows a screenshot (figure 6.6) from a decathlon game that has been "fixed" to show that Caitlyn Jenner was the winner of the 1976 Olympic decathlon ("HackFest | KansasFest" n.d.). A transgender woman, Jenner won the decathlon as Bruce Jenner; following her transition, Jenner was famously photographed by Annie Leibovitz for the July 2015 cover of *Vanity Fair* (Bissinger 2015).

The 2017 Catalan re-patching of *Summer Games II* (1985) and the 2015 "fix" of a 1980s decathlon game demonstrate that 8-bit game culture is a living culture. That users are continuing to do new things with older game sources suggests that games of the micro era have become a shared cultural inheritance. They not only carry rich resonances but are also marked by a sense of collective cultural ownership. Appropriations and cultural interventions utilizing the historic game corpus mark a moment when vintage games have become so culturally significant—so much a part of a shared and accepted set of digital cultural references—that they are seen as fit and proper vehicles for expressing current-day political struggles. Together with the persistence of practice on legacy computers, such reuse suggests that a charged moment in the appreciation of the microcomputer is upon us, a moment in which we might be able to examine the history and significance of our relationship with the microcomputer with greater nuance. At least that is the promise I see. In chapter 7, I highlight the significance of the homebrew case study for existing disciplines and point out where there is potential to open up new research directions.

7 New Directions

This book has addressed 1980s microcomputing, specifically the practices of users as they learned to program and developed games at home. Today, personal computing is ubiquitous, with software—frequently in the form of digital apps—used for an astonishing array of purposes, both mundane and important, from choosing a restaurant, to real-time monitoring of vital signs, to finding a partner. Digitality has arguably become part of the human condition. This wasn't always the case. Micros were the first experience of hands-on digital computing for many. Anyone interested and with the means could purchase a computer for as little as $300. The 1980s was also when many aspects of living transitioned from analog to digital, whether in the field of banking (e.g., the introduction of automated teller machines), entertainment (e.g., digital special effects in film), or information services (e.g., teletext). Approaching the moment when digitality was new offers the chance to consider whether we are asking the right questions of the computer's adoption and what new perspectives might emerge, both on this period and on the contemporary experience of digitality.

The reception of microcomputers in the 1980s was highly experimental, because while there were preconceptions about what a computer was, there was little in the way of commercial software available for many platforms. This meant that answers to the question of what computers were good for were quite open, and, for a time, it was left to the ingenuity of users, who developed a range of uses for computers by developing software. Playing and writing games figured prominently among such uses.

The development of games at home was not unique to 1980s Australia and New Zealand, but structural factors give these locales distinct features. For instance, it is often assumed that people progressed to micros from larger computer systems, but that was not the case among my informants.[1] Most adults and children in Australia and New Zealand hadn't played games on PDPs or minicomputers, as these were not common outside universities and corporate computing contexts,[2] and very few schools had any computer access prior to micros. My informants were more likely to have had exposure to arcade games or games on other micros than on a minicomputer.

While Australia and New Zealand were distant from the perceived centers of game development in the 1980s—typically thought to be the US and Japan—there was a considerable amount of game development occurring. Both nations had commercial game industries during the 1980s, though of different kinds. In New Zealand, many local game arcade and console manufacturers were trying their luck in the marketplace when microcomputers appeared (Swalwell and Davidson 2016; Swalwell 2015). Meanwhile, in Australia, several game development studios got a start in the 1980s, including Beam Software, Microforte, and Strategic Studies Group. In the Play It Again project, we documented more than 900 game titles—published and unpublished—that were written for microcomputers in the two countries (700+ in Australia and 200+ in New Zealand), and there are probably more that we missed. The nascent industry provided a backdrop to the experiences of ordinary people writing software with which this book has been concerned. My intention, however, has been to ensure that the written histories of the micro era include some mention of the actors, sites, technologies, and products away from both the perceived centers of production and the "official" industry.

This study is explicitly a user history, inspired by the scholarship of Michel de Certeau and his collaborators Luce Giard and Pierre Mayol, initially through their insight that we know little about the uses that people make of things and secondly that users and consumers are makers and producers of culture, a "perspective reversal" that "displac[es] attention from the supposed passive consumption of received products to anonymous creation, born of the unconventional practice of these products' use" (Giard 1998a, xvii). De Certeau and his collaborators' attention to ordinary people, everyday knowledge, and cultural practice, together with the operations that bestow legitimacy and cultural value, provide a theoretical toolkit that was ripe for application to microcomputing and the beginnings of vernacular digitality.

This theory from the 1980s was not only helpful for unpacking the homebrew case study; the lines of influence run in both directions. One

New Directions

of the central questions I set out to answer in this book is what a study of homebrew game development might contribute to media, cultural, and audience studies. Specifically, what are the continuities and discontinuities between homebrew gaming and the longer tradition in cultural studies and cognate areas of theorizing consumption and audience activity? In chapter 5, I observed the prominence of building, hacking, and fixing in early homebrew computing and game culture, and that these practices are almost completely absent in the scholarship on user creation and production. Tinkering with technology has a different lineage than much extant scholarship, but such practices are not necessarily antithetical. The de Certeauian theoretical framework of user practices I have deployed provides a way to write the hardware hacking of ordinary users back into the narrative of users as producers.

The de Certeauian framework has also enabled me to broach some larger questions. These include: What part might "ordinary culture" play in a history of computing and software? What is vernacular digitality today? Relatedly, what new areas of research might such practice presage? And finally, what implications does vernacular digitality in the micro era hold for born digital heritage more generally? While such questions are beyond the scope of this book, they point to emergent topics of research and nascent areas of scholarship and practice.

Ordinary Culture in the History of Computing

In chapter 1, I cited Haigh's claim that the lack of attention to micros results from generational lag. I suspect there are several other reasons for computer historians' reluctance to attend to micros. While the temporal closeness of the period of popular computing deters some ("historians generally avoid writing about recent events on which they lack a proper perspective"; see Campbell-Kelly and Aspray 2004, 207) and the field's focus to date on business and industry histories is also a factor, I suspect that another impediment is a discomfort with the everyday. While there are exceptions, in the main, computer historians have seemed reluctant to reconstruct the lives of ordinary people.

Throughout this book, I have argued that the history of homebrew is a key moment in the vernacular reception of microcomputers. The domestic, the vernacular, and the popular reception and consumption of computers are subjects that belong centrally within histories of computing. Embracing vernacular use and ordinary culture in a critical way promises to enrich the field of computer history, which is well placed to consider the ways in which legitimized and delegitimized—"official" and "unofficial"—are imbricated and "enrich each other" (de Certeau and Giard 1997, 105). Legitimacy changes over time, and there are indications that some aspects of the positioning of vernacular digitality are changing. For example, Joy Rankin's recent book *A People's History of Computing in the United States* is "a history from the user up," studying "computers for ordinary people" from the decades prior to microcomputers' appearance (Rankin 2018, 10).

Vernacular Digitality Today

Change in the treatment of vernacular digitality and computing bodes well not only for the future of computer and game history but also for the emergence and development of new areas of scholarship. History exists in a dynamic relationship with the present. The history of homebrew game development I have presented in this book offers new perspectives on and raises important questions about what it is to engage with a computer, including in the current moment. I look with a doubled glance—both looking from the contemporary historical moment to some 1980s antecedents and wondering at what practices the energies of the microcomputer era seeded. This is not an easy prehistory of the present. Rather, it has been guided by the conviction that we should be able to hold the historical and the contemporary together, to encourage a dialogue between the different historical moments, recognizing for example that "the past actively exists in the present" (Foucault 1984, 81). Equally, we need to consider whether we are asking the right questions of the current moment. Such an approach is consistent with media archaeology, where one searches for "unnoticed continuities and ruptures" (Huhtamo and Parikka 2011, 3), "excavati[ng] media-cultural evidence for clues about neglected, misrepresented, and/or suppressed aspects of both media's past(s) and their present" (Huhtamo 2011, 28).

It is my hope that the concept of vernacular digitality outlined in this study might be one that others find useful, so that questions about vernacular digitality in other moments might be ventured. What are the characteristics and textures of vernacular digitality now? How have our ways of using computers for ends not productivity related developed and changed over the decades? Are users still able to find joy in creating with computers, given they are such a part of the workaday world? Many are engaging in hobbyist pursuits, from high tech to traditional. What questions should we be asking about such activities? If relations between hobbies and paid employment are changing, how so?

In the midst of such concerns, the outlines of what I think of as a new subfield of hacking and tinkering studies are already discernible. At a time when digital media are pervasive and proprietary hardware is increasingly "black boxed" and locked down, there is significant pushback against the closed nature of computer environments and other consumer goods. Scholars are taking tinkering and hacking seriously, along with the urgent legal questions that modding and the circumvention of digital rights management (DRM) and technological protection mechanisms (TPMs) raise (Schulz and Wagner 2008; Wilson 2017; Gillespie 2009). As Andrew "bunnie" Huang—author of *Hacking the Xbox: An Introduction to Reverse Engineering* (Huang 2003) and many open hardware projects—is credited as saying, "If you can't hack it, you don't own it."

In the coming years, I anticipate the growth of scholarship around both historical and contemporary user practices, including hacking and tinkering. The public certainly intuits connections between what home coders made in the 1980s and the outputs of the do-it-yourself (DIY) movements now, and there is increased awareness of such practices, as some makers and tinkerers move out of their homes into shared spaces such as those auspiced by the maker and men's shed movements (Foege 2013). In Australia, Mark Thomson is a key artist-practitioner in this in-between space, with his whimsically named Institute of Backyard Studies, the motto of which is "I tinker, therefore I am" (Thomson 2002, 2007). But there are also new politics in play as the open hardware movement in computing has expanded into the right to repair movement. Some are radically scaling up their efforts, such as iFixit's database of medical repair manuals, which is taking the fight to corporations (Purdy 2020), and Adafruit, which was created to teach people how to hack and build electronics but pivoted during the Covid-19 pandemic to "building and shipping goods like face shields, sensors for medical devices, digital thermometers, oximeters for taking finger pulses, and thermal cameras for fever screening" (Chan 2020). If, as Douglas notes in her study of ham radio, the tinkerer's ethic was bound up with "key elements of twentieth-century masculinity—the insistence upon mastering technology, the refusal to defer to the expertise of others, [and] the invention of oneself by designing machines" (Douglas 1999, 328)—then these developments invite inquiry into the contemporary construction of gender in such activity, among other dimensions. Scholars in the new subfield of hacking and tinkering studies in all likelihood will look back not only to antecedents—building on the work of pioneers of tinkering (Takahashi 2000; Douglas 1992; Franz 2005)—but also to other cognate hybrid fields of hobbyist computing, digital fabrication, hacking, fixing, modding, and others. And they will draw on other perspectives that demonstrate interest in the interaction of hardware and software, such as platform studies. As Nick Montford and Ian Bogost write, "We believe it is time for those of us in the humanities to seriously consider the lowest level of computing systems and to understand how these systems relate to culture and creativity" (Montfort and Bogost 2009, vii).

Digital Heritage Now

Another field for which vernacular digitality holds strong significance is cultural heritage. As early as 2001, Henry Lowood—a pioneer of software history—articulated a rationale for collecting and preserving software, writing that

> the broader social and cultural impact of computing will revolutionize (if it has not already) all cultural and scholarly production. It follows that historians (not just of software and computing) will need to consider the implications of this change, and they will not be able to do it without access to our software technology and what we did with it. . . .
>
> Historians of software clearly will have to venture into every niche, nook, and cranny of society in ways that will separate their work from the work of other historians of science and technology. (Lowood 2001, 148–149)

Whenever I give a talk and include this quotation, people instantly get it: Lowood succinctly captures a compelling reason for why software matters, providing newcomers to the field of software heritage and preservation with a way to grasp the problems future historians of computing and software will face. However, I am starting to suspect it is time to revisit the foundations of the rationale for collecting in order to ensure that it explicitly goes beyond the scholar's cause, for one of the questions this homebrew study begs is surely that if we have neglected homebrew up until now, then what other practices and artifacts of ordinary computer culture have also been overlooked?

The reuse of software artifacts referenced in chapter 6 demonstrates not just public interest but also the public's *stake* in software heritage. People are locating personal and political meaning in digital cultural artifacts and using these artifacts to express solidarity with others. Appropriations and cultural interventions utilizing historic game software indicate the depth of vintage games' cultural resonance. The reuse examples highlight the nimbleness and dynamism of everyday digital culture, which exceed the rather polite terms in which software heritage has been articulated to date. Significantly, the games have been made available by unofficial archivists (crackers and fan preservationists) and are hosted by the Internet Archive, a body that explicitly foregrounds access and reuse, not just by scholars but by everyone.

That thirty-year-old microcomputer game titles are becoming vehicles for contemporary political expression underlines the need to ensure both that conceptions of digital cultural heritage are adequate to encompass such imaginative and innervative reuse and that collection policies are fit for the purpose.[3] Some years ago, I was playing around with definitions of digital heritage, wondering whether there was already a definition that captured my thinking on the subject. I found UNESCO's definition to be very focused on the urgency of safeguarding artifacts from loss. These are of course important considerations, but they don't address the issue of what makes people reach for a 1980s game when they are searching for an appropriate form for political expression. I found little that articulated the significance of digital heritage and how this might be different from other forms of heritage (Swalwell and de Vries 2013). Encouraged by an archaeological colleague, I decided to take some liberties with the Burra Charter, an International Council on Monuments and Sites (ICOMOS) document used to assess the heritage significance of place in Australia. My form of the charter reads: "Items of digital cultural heritage enrich people's lives, often providing a deep and inspirational sense of connection to self, others and community, to the past and to lived experience. They are historical records that are important as tangible expressions of identity and experience. Digital heritage items reflect the diversity of our communities, telling us about who we are and the past that has formed us, and about society and cultures. Items of digital cultural heritage are irreplaceable and precious. They must be conserved for present and future generations" (adapted from Australian ICOMOS 1979, 1).

The definition is not perfect, and in the case of games, for instance, there are some important caveats around the extent to which they "reflect the diversity of our communities." But this definition at least has the advantage of capturing aspects such as the sense of connection, of shared community, inspiring and expressing identity and experience that are exemplified in the examples of the Catalan hack of *Summer Games II* and "Olympic Decathlon fix" (2015) cited earlier.

If we accept that vernacular digital cultural heritage matters because it provides a sense of connection, a means of inspiring and expressing identity, then there are implications for collecting digital culture in the current moment. What are the contemporary equivalents of 1980s homebrew? What forms of the digital offer people opportunities to practice a "modest inventiveness," bearing in mind that the products of such activity might not be highly visible? I do not yet know the answers to these questions, but they need to be asked, repeatedly. If we have managed to overlook the ordinary culture of homebrew creation for microcomputers—because such practices were located in domestic space and therefore not obvious, because the activity was private and people didn't necessarily talk about it or it was derided, and we have perhaps also lacked a discourse in which to describe and therefore grasp the significance of this activity—then what else might we have missed? Cultural heritage professionals need to seek answers to such questions. Rather than eschewing unofficial culture, they need to look around for its traces, which, as de Certeau and his collaborators remind us, enrich official culture and vice versa. Both need to be collected if we are to be able to reflect on and study their interaction in the future.

Ordinary or vernacular digital culture and practice in the present and the recent past is at risk of being dismissed as not sufficiently important not only because of its unofficial status but also because of its digitality. Whereas a curator of ephemera (literally meaning "of the day") could usually count on material objects surviving years (or even decades) of benign neglect until they made it into a collection, digital objects do not always survive for very long without active intervention. My current research into various creative uses of microcomputers in the 1980s has yielded many examples of software and hardware dependencies that make artifacts challenging to reconstruct. This means that the timeframe in which collecting decisions need to be made is greatly condensed. The time to do such collecting is while the period remains within living memory, preferably as close as possible to the time of creation.

Two contemporary examples illustrate the dynamics of loss and retention of born digital artifacts. When I began writing this chapter, the microblogging and social networking website Tumblr announced that it would ban adult content. To demonstrate its responsiveness to child exploitation

New Directions 179

material appearing on its platform, the company would be deleting any content deemed not suitable for work, giving users a month to download or move their stuff. As I finish my edits, Melbourne—the city I live in—is shut down because of Covid-19, along with much of the rest of the world. During the shutdown, archivists from Australian and New Zealand collecting institutions have been discussing their approaches to collecting the digital ephemera of the pandemic on the local digital preservation network Aus-Preserves. These two examples demonstrate both the changing valuation of the vernacular and custodians' foresight and imagination in collecting and stewarding contemporary digital artifacts. Loss may be part of the story of twenty-first-century platform capitalism, but there is also evidence that institutions are responding and collecting everyday digital culture.

Nevertheless, if we think that our cultural institutions ought to include ephemeral records of the everyday, of unofficial and vernacular culture as well as official creation, then we need to articulate this clearly and explicitly task the custodians of these collections—our curators, librarians, and archivists—with this work and equip and resource them to do it. Heritage professionals face a number of challenges in collecting and stewarding digital objects, as their preservation requirements are quite different from those for material objects. Existing practices and protocols need to be adapted. It is fitting if not ideal that enthusiasts and other unofficial archivists have to date often been the ones leading the charge in caring for much digital cultural heritage, as they recognize its fragility. A measure of heritage professionals' success will probably be the extent to which they are willing and able to collaborate with unofficial archivists.

As heritage professionals undertake this work, this account of homebrew developers' practices and perspectives may be salutary to consider. I got to these informants—de Certeau's "ordinary man"—while they were still able to recount their stories, even though many of their digital artifacts are long gone. As Lowood said in 2015, "I think the thing to keep in mind is that in 2050, there will not be people around who went through the digitisation of daily life. The role of the documentation, the role of the software library, the role particularly of the historical documentation will be to represent to people . . . to give people something to understand what those transformations [were] like" (Lowood in Young 2015).

At present, people who remember the 1980s micro era are still alive and, as I have emphasized, the culture around 8-bit microcomputing is a living culture. While 8-bit microcomputing practices might not have been very visible to outsiders during the 1980s, the advent of the internet has made it possible for contemporary enthusiast communities to emerge on a different and much more connected scale. There are vintage computer festivals, conventions, and less formal face-to-face get-togethers, often organized around particular platforms. Individually and collectively, enthusiasts organize and undertake some remarkable projects. These are communities to which curators and archivists can and ought to turn for assistance.

Notes

1 Introduction

1. While undoubtedly an important scene in computer history, with some practices that overlap those on which I focus, the history of this group is reasonably well known and frequently reprised. By contrast, this book presents a study of the little-known practices of home coders developing games from the late 1970s through the 1980s, from a region not often considered in game history, Australasia. One thing that the famed Homebrew Computer Club connection makes clear is the connection to computer hardware hacking, a topic I address in chapter 5.

2. Excavating the hidden histories of homebrew not only adds texture and richness to a history of computing and digital games at the moment when digitality became tangible for everyday users but also potentially provides a prehistory of indie game development. In a number of places, Anthropy and others seem to anticipate that 1980s homebrew provides a historical context for the contemporary retro homebrew and indie scenes.

3. Graeme Kirkpatrick's argument that evaluative criteria for games were still being formed at this time is worth remembering (Kirkpatrick 2017).

4. Indeed, Tom Lean makes the point that much of the output of some game companies—such as Imagine—was "of middling quality." What Imagine had, he argues, was a "genius for publicity" (Lean 2016, 180–182).

5. King and Borland variously write that programmers and game developers were often amateurs in the business world (King and Borland 2003, 44) and that as Richard Garriott was approached to commercialize Akalabeth, "The early hackers' day was already coming to an end ... as ... hobbyists left their basements for the lure of business profits" (King and Borland 2003, 40).

6. Exceptions include King and Borland (2003), Rehak (2008), Friedman (2005), and Newman (2017) and more recent titles in platform studies (Maher 2012; Gazzard 2016). Self-published sources on microcomputing in the US include Savetz (2012),

Scott (2005), and O'Hara (2006), with a recent addition from Australia focused on Macintosh games (Moss 2018).

7. Graeme Kirkpatrick has somewhat mischievously claimed that users in the US were not actually using their micros as they were in the UK—"they were bought but stashed away in cupboards" (Kirkpatrick 2015, 9). This is a view Kirkpatrick attributes to "Dan Gutman, *CU*'s [*Commodore User*'s] 'US correspondent,' who reported that American parents, like their peers in Britain and elsewhere, had bought home computers when the market in TV gaming consoles crashed, partly for their anticipated educational benefits, but then discovered that, 'when you bring it [the computer] home, plug it in and turn it on, *it doesn't do anything!*'" While I suspect Kirkpatrick's tongue is well and truly in his cheek in citing this view, it is fair to say that—with the exception of some iconic individuals whose homebrew origins have received attention, such as Richard Garriott/Lord British and Ken and Roberta Williams (King and Borland 2003; Donovan 2010)—homebrew microcomputer game development and use in the US market is an underresearched topic. However, to suggest that US users did not use their micros to program is a stretch, particularly given ample evidence of code listings in magazines such as *Byte*, *Dr Dobbs Journal*, *Softtalk*, and others.

8. As does de Certeau's use of the term *poiesis*—from the Greek *poiein*, meaning to create, invent, or generate.

9. Chapters in my forthcoming edited anthology *Game History and the Local* will add to the available scholarship, further theorizing the significance of locale.

10. I conceive of local game scholarship as a form of microhistorical endeavor (Ginzburg 1992; Levi 2001) but also attempt to find rapprochement between micro and macro historical perspectives.

11. Germaine Halegoua writes on this issue with considerable nuance, citing Gerard Goggin's critique of an overreliance on political economic approaches to global media and the tendency to produce "centers" and "peripheries" or "right places" and "wrong places." Halegoua argues that "in studying these 'other' or 'wrong' places we should aim to understand their particularities within global and local flows rather than simply reify them as marginalized or peripheral places within global networks" (Halegoua 2016, 6–7).

12. Carlo Ginzburg's study *The Cheese & the Worms: The Cosmos of a Sixteenth-Century Miller* uses the archives created around the inquisition of an Italian miller, Domenico Scandella, also called Menocchio, who is tried by the Catholic Church for heresy, tortured, and burned at the stake (Ginzburg 1992, xiii). Detailed court records exist because the Church required that a transcript be made. Menocchio—a self-taught but intellectually voracious reader—developed "his own startlingly eccentric cosmology" at odds with Church doctrine. His crime was holding heterodox views. Ginzburg describes Menocchio's views as "heterodox" in both general

and precise ways. Generally, his views differed from the then accepted cosmology (19), but his second use is more precise: Menocchio's "heterodox opinions" go against the Church's orthodoxy (21).

13. Amateur computer historians are already asking these questions. For instance, Rob O'Hara writes, "It both surprises and amazes me that more effort has not gone into documenting the BBS era" (O'Hara 2006). Some are not waiting and are documenting the history themselves, interviewing key informants and collecting documentation and software, as in the case of Kevin Savetz, who has "done more than 250 oral history interviews with people involved with the early home computer industry" for his Atari 8-bit themed podcast "Antic."

14. Editors Gerard Alberts and Ruth Oldenziel introduce their *Hacking Europe* anthology by deferring to "American dominance in computing" and "US cultural, political, and technological dominance," a narrative that seems to derive from both a Cold War legacy of computing and an identification that the personal computer industry was American. Their focus is firmly on how European users appropriated the microcomputer, which is taken as being American (Alberts and Oldenziel 2014, 4, 5, 7). By contrast, the Australasian context is far less deferential toward the US, probably deriving from a combination of factors: the historic relationship with England (e.g., Stuckey 2013a), proximity to Asia, homegrown innovation, and interest in American markets.

15. I first started thinking about user making in de Certeau's terms when I was researching 1990s LAN gaming practices for my PhD, but while homebrew games and mods are both user-produced cultural artifacts, homebrew is quite distinct from modding: 1980s homebrew predates modding (which arguably began with *Doom* in 1993) and emerged in a completely different context, at a time when the game industry barely existed. Homebrew involved developing games from scratch, whereas modding is about changing *some aspect of* an existing game (an avatar's skin, a texture, a map), though full conversions are also made (Champion 2012b, 12–13). Yet another difference is found in the tools that were available to modders and to homebrew game developers.

Creating 1980s homebrew games is not modding, though other practices of the micro era—such as cracking—might be thought of as antecedents to modding.

16. Interestingly, this way of interviewing bears quite a strong similarity to the way de Certeau's collaborators Luce Giard and Pierre Mayol collected their material (for Mayol ethnographically and for Giard by interviews conducted by Marie Ferrier). It was hoped that the interviews would elicit "everything that usually remains unsaid about knacks for doing things, decisions, and feelings that silently preside at the accomplishment of everyday practices." Commenting on Ferrier's technique, Giard writes that she "discovered how to strike up with her female interlocutors conversations of a remarkable freedom, rich in unexpected information" (Giard 1998a, xxviii).

2 Discourses about Microcomputers

1. This chapter is based on reviews of an extensive range of Australian primary source materials, including general and specialist computer newspapers (*Sydney Morning Herald, Australian Microcomputer Magazine, Australian Computer Weekly, Pacific Computer Weekly,* and *Australian Microcomputerworld*), computer magazines (*Your Computer: Magazine for Business and Pleasure, Online: The Microbee Owner's Journal, Australian Commodore Review, Australian Commodore and Amiga Review, Australian Apple Review,* and *Australian Home Computer GEM*), and early code books. Searches of New Zealand newspapers and computer magazines (*Computer Input, Bits and Bytes,* and *Sega Computer*) have also been undertaken in the context of a broader history of gaming (e.g., moral panic around arcades). While there are many similarities and the New Zealand material broadly supports the arguments presented here, such sources are only referred to selectively in this chapter, for instance in the discussion of *Bits and Bytes* and *Sega Computer*.

2. Ironmonger, Lloyd-Smith, and Soupourmas extrapolate ownership figures from Roy Morgan surveys of consumer purchasing choices from 1985 to 1995.

3. Computer penetration and usage ("households in which a computer is frequently used") in Australia seems to have been fairly similar to that of other countries internationally, although the ABS notes the difficulty of making comparisons. There was very little data available for comparable reference periods, and what was available came from a variety of sources. Frequent use in Australia (31% in 1996) was slightly lower than in the US (34% in 1996) and the UK (33% of households owned computers in 1997) but was higher than in Canada (29% in 1995) and Finland (25% in 1995). The Netherlands appears to be an outlier at 43% (in 1996) (Australian Bureau of Statistics 1997, 5–6). It is significant that in 1996 the vast majority of computers in Australian households were owned by members of the household (81%, as opposed to 15% owned by a home-based business or an employer) (Ironmonger, Lloyd-Smith, and Soupourmas 2000). Unfortunately, Ironmonger, Lloyd-Smith, and Soupourmas don't break down their earlier figures by home or business use.

4. Figures from subsequent studies of computer ownership indicate that preparing family budgets was the main use of the computer in only 1.5% of Australian households where a computer was frequently used (Australian Bureau of Statistics 1994, 6), while keeping personal or family records was nominated by only 6% of users as the activity on which most time was spent (Australian Bureau of Statistics 1997, 47).

5. "The Commodore 64C Family Pack [included] 5 software programs for games, entertainment, education and finance management: *Wizard of Wor, International Soccer, Visible Solar System, Magic Desk, Financial Advisor.*"

6. These doubts as to computers' usefulness in the home persisted well into the 1990s. In its 1996 study on "Household Use of Information Technology," the ABS reported that "of the 4.4 million households which did not have computing

facilities, 40% gave 'no use for one' as the main reason for not having a computer, 30% said 'costs are too high' and 14% said 'no one in household interested in computer'" (Australian Bureau of Statistics 1997, 6).

7. Ironically, the editor of this publication, Andrew Farrell, had noted only a couple of years earlier that "smaller computers like mine don't have a lot of use in the business world because they're so limited" (Farrell cited in Filatoff 1983, 30).

8. The computers reviewed were the Amstrad CPC664, the John Sands Sega, the Sinclair QL, and the Microbee "Computer in a Book."

9. The provenance of these sales figures is not given, but they are likely to be Northern Hemisphere figures.

10. Such self-learning might also be supplemented by computer courses, such as those run by Computer Seminars (Australia) from their offices in Campbell Parade, Bondi. This company also allowed students to hire a computer for $39 (plus a $60 refundable deposit) for the full length of their course, subject to availability. For $5 extra, they would even mail the computer, via certified mail (Paull and Kovac 1984).

11. BASIC was sometimes maligned by professionals as a crude and even "disgusting" language for amateurs, but, as Barnes noted at the time, "different styles or cultures of programming" exist, adding that, "It is clear that the style of programming of a member of a large team developing a 100,000 line program for a complex defence system is utterly different from that of the lone hobbyist amusing himself with a 100 line game" (Barnes 1982, 10).

12. "A great deal has been said and written about the 'computer revolution.' It is here, and it is affecting all our lives in ways we don't even know about yet. The younger you are, the more impact computers will have on you—since you presumably have more years ahead to live in a world with more and more computers in it" (Ault 1983, v).

13. A further example of this claim is given by Street, the author of a ZX Spectrum guide, who wrote, "Programming a computer is a fulfilling activity. It presents a direct and compulsive challenge to our powers of organization, perseverance, and reasoning. The home computer is, of course, a many-faceted tool—with the right software it can entertain, inform, and act as an electronic filing cabinet—but it can also engage our wits and stamina in an attempt to make it carry out our particular will (Street 1983, 1).

14. Without discounting the prominence of game titles, it is worth noting that the software that was visible to people at the time depended on the circles they mixed in. The question of what other sorts of software might have been written in the 1980s but overlooked remains open. In chapter 6, I discuss some of the other genres of software (educational and "X-rated") that unofficial archivists are turning up for the Apple II.

15. Few of the local magazines were involved with circulation audits, so I have been unable to source circulation figures. Big name US and UK magazines (*Dr Dobbs*, *Byte*

magazine, and the UK's *CVG*, for instance) were available in Australasia, and some of my informants reported reading them. Local magazine writers clearly were also reading overseas publications and deriving story ideas and information from them. The local was thus inflected by, and imbricated with, the nonlocal. In one interesting example, a husband and wife team based in Broadmeadow, Sydney, republished the US TRS-80 Color Computer magazine *Rainbow* beginning in November 1987, with Falsoft's permission. Networks also extended across the Tasman Sea: Australian Sega SC3000 owners were keen readers of *Sega Computer* magazine, and newsletter exchanges between microcomputing clubs were also common.

16. The extent to which the Sega Users Club was actually a club remains an open question. For Grandstand, the main aim of the magazine—and its publishing of software—was always to generate interest around the Sega in order to move more stock. The Sega was a product to sell, much like any other. But to say that a club ethic was not uppermost in Grandstand's minds is not to say that users did not enjoy aspects of the club experience. As well as being an extremely effective strategy for creating a buzz around their product, the publishing of user group contacts in the magazine supported users in physically getting together.

17. Detective work by two New Zealand collectors—Aaron Wheeler and Michael Davidson—established that around three hundred software titles were published in New Zealand and Australia for the Sega SC3000 system alone (Swalwell 2009, 268).

18. See, for instance, Kirsten Haring's book on ham radio enthusiasts, *Ham Radio's Technical Culture*.

19. The acronym BASIC stands for Beginners All-purpose Symbolic Instruction Code. It was the language in which most hobbyists began to program.

3 Micro Users as Makers

1. While Friedman notes that there were other computers—"Following the success of the Altair, a number of other small computer manufacturers started up, selling to the growing hobbyist market"—he doesn't treat them or the programming these machines facilitated (Friedman 2005, 105).

2. Giard reflected on this in her chapter "Times and Places," noting that, "We . . . have discovered a certain echo in English-speaking countries right down to Australia, an echo in the disciplines of urban sociology, cultural anthropology, 'communication,' or in a new field, not yet recognized in France, cultural studies, a new way of writing the history and sociology of contemporary culture" (Giard 1998b, xlii).

3. While it is often understood that what users produce is cultural—this has been the dominant understanding in cultural and fan studies—Maker Faires also helpfully demonstrate that user production is also literal, as in the act of making a material artifact.

4. That sometimes one must make do because of scarcity is not at issue (de Certeau, Giard, and Mayol 1998, 43), but the reference usually is to there being a paucity of cultural options on offer. Jenkins writes that Fiske had a "crisis of faith, returning many times to the idea that the people lack the infrastructure and resources to sustain their own forms of cultural production" (Jenkins 2010, xxxii).

5. De Certeau was added as an author on a second edition of volume 2 of *The Practice of Everyday Life*, both to "salute his memory and to make visible his presence in this volume" and because of the inclusion of two of his sole-authored pieces and one coauthored with Giard, published after the publication of volume 1 (Giard 1998b, xlii).

6. What references there are tend to be to anonymous disciplinary mechanisms. Elsewhere, for instance, de Certeau mentions the historian's reliance on the computer and quantification and makes reference to the general turn to the number as the "chief index of historical truth" (Poster 1992, 98).

7. Anecdotally, even some people who claim to have bought and played a lot of computer games in the 1980s have reported never having come across a homebrew title.

8. Indeed, Timothy Tomasik makes clear that de Certeau was only credited for his preface to the first edition of volume 2, not as a coauthor of the book: "When Giard edited and published a revised, expanded edition of this volume in 1994, she and Mayol decided to include Certeau as a coauthor since they were adding one essay by Certeau and another that he had written with Giard" (Tomasik 2001, 519). Despite Giard having written a very clear introduction to volume 2 outlining the history of the research project, there is significant potential for confusion. This is compounded by the fact that she is not listed as an author of the book in which sections of the coauthored report appear (*The Capture of Speech and Other Political Writings*). For the sake of clarity, I reference this section of the book using both authors' names (de Certeau and Giard 1997).

9. Giard indicates that this was chapter 1 of the 1983 report (Giard 1997, xviii). On my reading, material seems to have been drawn from a range of chapters (de Certeau and Giard 1997).

10. Most of the chapters appear in English translation (by Tom Conley) in part III of *The Capture of Speech and Other Political Writings* (de Certeau 1997).

11. This, I think, is the meaning de Certeau and Giard give to the slightly obscure term *operativity* or sometimes *operations*, or *ways of operating*. It is a loose synonym of *uses*, *practices*; for example, "This essay is part of a continuing investigation of the ways in which users—commonly assumed to be passive and guided by established rules—operate" (de Certeau 1984, xi).

12. At the time of our interview, John and Joseph White were creating games under the name Lexaloffle.

13. There are echoes here of the concept of *la perruque*, which de Certeau introduces as a "diversionary practice" where "the worker's own work is disguised as work for his employer" (de Certeau 1984, 24–25).

14. I acknowledge Wade's point that the term references earlier studies of bedroom culture (Wade 2016, 57).

4 The Games

1. Interviewing Symons, we noted that some of his books were published by Corgi and also by Addison-Wesley, the American educational publishers. According to Symons, Hartnell published under multiple aliases, "so he already was the most prolific author, but what people don't realise is he had a number of aliases as well. So Andrew Nelson was one of his that must have done 20 books, and it was Tim Hartnell. And the issue was that if he had a book with a publisher that slightly went onto the area of another book he wanted to write, then he had to have a different author and it had to be through a different publisher. So he'd use Interface Publications for his own stuff but then he'd go through Pitman, Corgi, Bantam, Penguin. . . . So they were spread around [so as] not to infringe" (In-person interview, April 23, 2015).

2. Elsewhere, I have argued for the importance of situating games relative to other visual media and their histories and observed that the boom in game studies has not yet led to assessments of digital games' wider relations to and significance for visual culture (Swalwell 2007).

3. Several scholars have published studies of microcomputer clones from their regions, and the variety of circumstances helpfully illustrates how cloning was simultaneously common yet took different guises. Gazzard notes that unlicensed "arcade clones" programmed at home were exceedingly common in the UK up until about the mid-1980s, and that as well as assisting with the nation's computer literacy push, they extended the experience of the arcade (Gazzard 2014). For Fassone, it is significant that the Italian clone *Camelli*—sold on tape at newsstands—uses the source code of its target *Attack of the Mutant Camels*, with only splash screens and titles being changed (though I would suggest this makes *Camelli* a crack rather than a clone); indeed, Fassone posits that the source was likely an Italian cracker (Fassone 2017). Meanwhile, *Pac-Man for the Vic-20* (1984) appeared as a source code listing in the Finnish magazine *MikroBitti* (Saarikoski, Suominen, and Reunanen 2017).

4. A company such as Nintendo, for instance, released some of its most famous games (e.g., *Donkey Kong*) on multiple systems—arcade, Game & Watch, Famicom/NES, through to the Wii—but also third-party hardware, such as Atari 2600/VCS, 7800, Intellivision, Amstrad, and others. Nintendo continues to rerelease their "classic" game titles on contemporary consoles—such as variations of *Mario* on the Nintendo 3DS XL.

5. We also see it in such contemporary casual games as Hipster Whale's *Crossy Road* (2014), which is also—in one sense—*Frogger* (1981). It is not surprising, then, that *Crossy Road* should itself receive the "demake" treatment, something I discuss in chapter 6.

6. Alongside amateur adaptations, porting was also serious business. Helen Stuckey's research into Beam Software's development of *The Hobbit*—first published for the Sinclair ZX Spectrum in 1982—is a case in point (Stuckey 2014). The game's success led to its being published on many popular home microcomputer systems, including the Commodore 64, BBC Micro, Dragon 32, MSX, Amstrad CPC, Apple II, Macintosh, and PC. While a number of these used the graphics developed for the Spectrum version, including a 1983 Commodore 64 version, many were also based on a later (1985) Commodore 64 version with enhanced graphics and the addition of music. While each iteration of the game uses the same parser and source code, each platform's individual constraints transformed the look and feel of the game.

Porting continued into at least the mid-1990s, when Nick Westgate and Sean Fausett were employed to port English educational software from the BBC computer to the Apple II for Softime (NZ) Limited.

7. The Microbee and Sorcerer had very similar architecture.

8. Quite a few of my informants went on to work in the computer, software, or game industries (Simon Armstrong, Vaughan Clarkson, Matthew Hall, Nickolas Marentes, Katharine Neil, John Passfield, Darryll Reynolds, Mark Sibly, Bob Smith, Andrew Stephen, Arthur Streeter, Ross Symons, and John White). Respondents to a recent survey by HackerRank suggest that the 1980s were remarkable in that people learned to program from a very young age. Examining the current composition of programmers in the workforce, the report notes that 12.2% of programmers between the ages of 35 and 44 started coding between the ages of 5 and 10, compared with 68.2% of programmers 18–24 years old, who started coding between the ages of 16 and 20 (Hughes 2018). Of countries where there were at least one hundred responses to the survey, the UK topped the list of countries with the highest share of developers coding between ages 5 and 10 with 10.7%, followed closely by Australia with 10.3% (HackerRank 2018).

9. Marentes relates that he was looking around for something new and innovative and found inspiration in an arcade game he'd read about that used atomic particles: "Each particle has certain characteristics and you've got to learn the characteristics of the enemy particles, learn the characteristics of your particles, and then you have a certain goal that you've got to strive for and work within that framework." He believes that it was "too abstract for a lot of people."

10. To me, this turn of phrase suggests a production without accruing capital—something that suits many of my informants' curiosity-driven game development activities though not all—but de Certeau clarifies that it means "without taking control over time" (de Certeau 1984, xx), a rather more abstract concept.

11. Thanks to Laine Nooney for alerting me to the Broderbund advertisement.

12. On occasion, Streeter bought games from other people and incorporated them into his compilations. He recalls, "What would happen occasionally is that someone who had bought one of my games would send me a game that they'd written themselves and some of those, they were typically young fellows, you know, high school students usually, and some of those I encouraged and paid them money for their copyright and incorporated their games into Street Games."

13. De Simone convinced the main creditor, Westpac bank, that he would be able to turn the company around within three years, believing that the computer's strong position in the school market meant that "Microbee is in a better position than anyone else to attract [the home market], because 500,000 school children turn them on each day" (McBeth 1988).

14. I am also a little mystified as to what the stakes are in claiming such a decline in practice.

15. For Lean, "hobbyist" seems to be synonymous with building one's own computer, whereas I would argue that those who bought and used a readymade computer to write games for themselves at home were also hobbyists. The different usage is not important except to say that such users were very much engaged in production, so it is not enough to designate them as simply "users" of a "mass market" machine (Lean 2016, 54, 68).

16. One informant, Passfield, moved out of his self-confessed "obsession" with coding for a time for fear of appearing too "nerdy," a stereotype he was sensitive to, but then he came back to it.

5 Hardware Hacking and Electronics

1. It seems likely that the number of hobbyists who were hacking hardware in the 1980s was relatively low, though probably higher compared to the general computing population than it is now. A 2008 survey, for instance, found that around 20% of respondents to a survey of Xbox hacking sites were what they called "user innovators," with the rest being "adopters" (Schulz and Wagner 2008, 413).

2. Patryk Wasiak, for instance, writes that the high cost and shortage of peripherals meant that hardware modifications were popular in Poland (Wasiak 2014, 138).

3. As Kristen Haring writes of so-called hams, "radio hobbyists were people who took up technology for leisure, forming their own 'technical culture'" (Haring 2007, xv).

4. Looking in the opposite direction, as "geeks with an adventurous side, who could be counted on to solve (and cause, sometimes) electrical problems," Haring argues that hams were "precursors to computer hackers" (Haring 2007, xv).

5. Besides Jim [Jamieson] Rowe VK2ZLO, the current staff of *Silicon Chip* magazine includes Ross Tester VK2KRT and Rodney Champness VK3UG.

6. I have heard anecdotally of a New Zealand pirate radio station that broadcast software. Each broadcast would be prefaced by a signal (e.g., three beeps) that told the listener to press record on their cassette recorder at home (Mark Williams, personal communication, July 7, 2015). This distribution method was much better established in Europe, particularly Eastern Europe (Beregi 2015; Jakić 2014; Veraart 2014).

7. According to a family member (who posted the John Heilborn columns at http://computerhistorycolumn.wordpress.com/about/), Heilborn was a member of the Homebrew Computer Club.

8. The question is sometimes asked whether homebrew isn't the same as modding, to which the answer must be "no." Attempting to apply 1990s modding practice to 1980s homebrew game development is anachronistic. Case modding is, however, plausibly descended from the practices I am discussing.

9. This is clear when Giard reflects on the reception of volume 2, that it was less read by the American public, but "we . . . have discovered a certain echo in English-speaking countries right down to Australia, an echo in the disciplines of urban sociology, cultural anthropology, 'communication,' or in a new field, not yet recognized in France, cultural studies, a new way of writing the history and sociology of contemporary culture" (Giard 1998b, xlii).

10. After having enjoyed success with his use of de Certeau's notion of poaching, Jenkins declares—in the name of theoretical renewal—that in writing "Interactive Audiences," he set himself the goal of "writ[ing] about fans without once mentioning Michel de Certeau" (Jenkins 2006, 134).

11. The growth in the field of web history is very welcome. Fan and otherwise homemade websites are a particular focus of Dragan Espenschied and Olia Lialina, who host what they call "The GeoCities Research Institute" (http://blog.geocities.institute/). See their book on the "vernacular web," Niels Brügger's book *The Archived Web: Doing History in the Digital Age*, the scholarly journal *Internet Histories*, and the conference series "The Web That Was" (Brügger 2018; Lialina and Espenschied 2009). I hope that as this nascent field develops it will gain traction in both media histories and fan studies.

12. Michele White, for instance, treats internet use as spectatorship, discussing this in terms of the gaze (White 2006). Such framings of practice are evidently not adequate to notice the types of user activities with electronics I've been outlining.

6 The Legacy of 1980s Homebrew

1. Giard's observations of the impact of professionalization on the home cook are germane here. She writes of the "tiny metal instruments" that exist "to give 'professional' perfection" to the dishes of the home cook, opining that it "is a pity because it is as if she has to mimic the production of a caterer or an industrial cookie factory in order to please her guests" (de Certeau, Giard, and Mayol 1998, 210).

2. The 8-bit aesthetic has undoubtedly undergone a revaluation from a time when perceptions were that two-dimensional blocky graphics looked "primitive." While some no doubt always loved the look of 8-bit graphics, it did suffer in the quest for the "holy grail" of photorealistic computer graphics before a renaissance of sorts began early in the new millennium, when it was realized that visually simple 2D games were well suited to the small screens of mobile telephones.

3. Smith is a prolific homebrew creator. At the time of writing, his creations include *Domin8tr1s* (2010); *Virus* (2010); *Boulder Logic* (2011); *Miner Man* (2011) and *Noir Shapes* (2012), both Xbox 360 demakes of Electric Wolf games; *Impact!* (2012); *One Little Ghost* (2012); *Ant Attack* (2013); *Quack* (2014), a *Flappy Bird* clone; *ZXagon* (2014); *Pandemic* (2014); *Rebound* (2014); and *U-Bend* (2015). And these are just Smith's game titles for the ZX81; he also counts several mobile and many ZX Spectrum games among his oeuvre. Though Smith sometimes gets asked to write for other old platforms, he has "a certain love . . . for the ZX81 because it was the first machine I had."

4. Smith told me that he likes "to try and be 'above board'" about making versions of other people's games, so he contacts the developers and includes copyright notices and rightsholder information on the loading screen of those games he has based on another title. He has often used the terms "conversion" and "remake" or "based upon" to describe his version of a game. In the case of ZXagon, the title screen reads: "ZXagon Based upon 'Super Hexagon' Copyright Terry Cavanagh 2012. This version released 2014. Code and Graphics by Bob Smith. This game is released with the full permission of the copyright holder and is for non-commercial use only" (Smith n.d.e).

5. The case of the demoscene is instructive to consider here. It is a significant microcomputing subculture that sometimes overlaps homebrew game development. Despite demoscene scholarship still being at a nascent stage (Reunanen and Silvast 2009; Hansen, Nørgård, and Halskov 2014), the scene has been comparatively better at self-archiving, quite probably because of its collective, community aspect. Several archives exist, including the "Got Papers?" site, which collects the scene's material heritage (Albert n.d.). In addition, campaigns are under way to have demoscenes inscribed onto UNESCO's list of Intangible Cultural Heritage through the Art of Coding campaign (Art of Coding n.d.), which nicely illustrates the shifting legitimacy accorded to informal computing practice.

6. I say "largely" because in 2019 Nintendo issued a takedown notice for a Commodore 64 port of *Super Mario Bros* (Orland 2019).

7. A video was embedded in the tweet a2_poet posted, "Current events in my homeland—Catalonia—and retrocomputing come together at last. Here my unreleased 30 yr old hack to *Summer Games II*."

7 New Directions

1. Swaine and Freiberger write that "games were nothing new to the early hobbyists who had played them on the big computer systems at their jobs, sometimes even loading games into memory on large time-sharing systems" (Swaine and Freiberger 2000, 165).

2. Some users of larger computer systems worked for companies such as Beam Software and Microforte, and Stephen Jones has documented an earlier period in Australian computer history that focuses on such systems (Jones 2011).

3. While innervation is often associated with decline and atrophy, I am using the term here in line with Walter Benjamin's understanding of innervation as a two way process, which also accords with usage in anatomy, to stimulate or supply with energy. I have elaborated this at length elsewhere, following Miriam Hansen (Swalwell 2002, 2008b; Hansen 1999).

Homebrew Software Cited

Title, developer(s) if known, date, and publisher (if published) are given.

Ant Attack, Bob Smith, 2013.

Auction Lots, ArComPro, 1986.

Bar File, ArComPro, 1985.

Beef Stud File, ArComPro, 1986.

Boulder Logic, Bob Smith, 2011.

Bunyip Adventure, Ross Williams, 1984, Grotnik Software.

Canberra Canberra, Dorothy Millard, date unknown.

Chilly Willy, John Passfield, 1984, Honeysoft.

City Lander, John Perry, 1984, Grandstand.

Compu-B (aka *Computer Betting*), John Schellens and James Roe, c.1984, Dreamcards.

Cordial Stall, Dean Hodgson, date unknown, Tandy.

Cosmic Bomber, Nickolas Marentes, 1982, Supersoft Software.

CroZXy Road, Bob Smith, 2015.

The Dare, Dorothy Millard, 1989, base7 Software.

Dinky Kong, Mark Sibly, 1984, Perspective Software; date unknown, Kiwi Computer.

Domin8tr1s, Bob Smith, 2010.

Donut Dilemma, Nickolas Marentes, 1984, Fun Division; 1986, Tandy.

Emu Joust, R. Sharples and G. Colmer, 1983, Honeysoft/Mytek.

Gate Crasher, Nickolas Marentes, 2000.

Genealogy, ArComPro, 1985.

The Gladiator, Nickolas Marentes, 1983, Fun Division.

Gloom, John Kowalski, 1996.

Gridfire, Vaughan Clarkson, 1983, Honeysoft.

Halloween Harry, John Passfield, 1985, Honeysoft.

Harboro, Dorothy Millard, 1990.

Harbour, John Perry, c. 1984, published as a listing in *Computer Input*.

Hoards of the Deep Realm, Vaughan Clarkson, 1985, Honeysoft.

Impact!, Bob Smith, 2012.

Jet Set Willy, Mathew Smith, 1983, Software Projects.

Jewels of Sancara Island, Matthew Hall, 1988.

Laser Hawk, Andrew Bradfield and Harvey Kong Tin, 1986, Red Rat.

Maths Invaders, Dean Hodgson, date unknown, Tandy.

Merlin, Lindsay R. Ford, Dreamcards, 1983.

Mozzie Zapper, Arthur Streeter, 1987, Street Games.

Neutroid, Nickolas Marentes, 1983, Fun Division.

Noir Shapes, Bob Smith, 2012.

Olympic Gold, ArComPro, 1984.

One Little Ghost, Bob Smith, 2012.

Pac-Man Tribute, Nickolas Marentes, 1997.

Pandemic, Bob Smith, 2014.

Pony Jamboree, ArComPro, 1984.

*Pop*Star Pilot*, Nickolas Marentes, 2016.

Psychotec, Lindsay R. Ford, Dreamcards, 1982.

Quack, Bob Smith, 2014.

Quizmaster, ArComPro, 1984.

Radio Operator's Logbook, ArComPro, 1986.

Rebound, Bob Smith, 2014.

Rupert Rythym, Nickolas Marentes, 1988, Tandy.

The Search for King Solomon's Mines, Darryll Reynolds, 1986, Softgold.

Homebrew Software Cited

Secret of Bastow Manor, Darryll Reynolds, 1983, Gameworx.

Sharemarket II, ArComPro, 1984.

Showjump, ArComPro, 1985.

Sirius 7, Cameron McKechnie, Rodney Smith, and Blair Zuppicich (Art Software), 1990, CRL.

Sorceror's Apprentice, Rodney Smith, Mark Sibly, Blair Zuppicich, and Cameron McKechnie (Art Software), 1990, CRL.

Space Intruders, Nickolas Marentes, 1988, Tandy.

Spelling, Dean Hodgson, date unknown, Tandy.

Squash Controller, ArComPro, 1984.

Stellar Odyssey, Nickolas Marentes, 1982, Fun Division.

Stellar Odyssey Part II, Nickolas Marentes, 1983, Fun Division.

Stranded, Bob Smith, 2005, Cronosoft.

Thermonuclear WarGames, Gameworx/Darryll Reynolds, 1984, Severn.

U-Bend, Bob Smith, 2015.

Virus, Bob Smith, 2010.

Warranty Recorder, ArComPro, 1984.

Weight Recorder, ArComPro, 1984.

ZXagon, Bob Smith, 2014.

Works Cited

"1983's Top Sellers." 1984. *Australian Microcomputer Magazine* 1 (10): 57–58.

4am. 2017a. "Passport." https://archive.org/details/Passport4am.

4am. 2017b. Twitter, October 31, 2017. https://twitter.com/a2_4am/status/925134895395074048.

4am. 2018. Twitter, August 26, 2018. https://twitter.com/a2_4am/status/1033722440562737153.

4am. 2019. "Apple II Library: The 4am Collection." https://archive.org/details/apple_ii_library_4am.

a2_poet. 2017. Twitter, October 26, 2017. https://twitter.com/a2_poet/status/923321967175983106.

a2_poet. 2019. "Sgii_catalonia_patch_to_4am_crack.Txt." https://ia800708.us.archive.org/23/items/a2_sugaiicat/sgii_catalonia_patch_to_4am_crack.txt.

Abbate, Janet. 2012. *Recoding Gender: Women's Changing Participation in Computing.* Cambridge, MA: MIT Press.

Ahl, David H. 2010. "Further Thoughts." In *Small Basic Computer Games—Small Basic Edition*, edited by David H. Ahl and Philip Conrod, 13–16. Maple Valley, WA: Biblebyte Books.

Albert, Gleb J. n.d. "Got Papers?" Scene.org. Accessed June 17, 2020. https://gotpapers.scene.org.

Alberts, Gerard, and Ruth Oldenziel, eds. 2014. *Hacking Europe: From Computer Cultures to Demoscenes.* London: Springer.

Allen, Robert C. 1985. *Speaking of Soap Operas.* Chapel Hill: University of North Carolina Press.

Alpers, Meryl. 2014. "'Can Our Kids Hack It with Computers?': Constructing Youth Hackers in Family Computing Magazines (1983–1987)." *International Journal of Communication* 8:673–698.

Anonymous. 1984. "Sega's Young Programmers: Today New Zealand, Tomorrow the World." *Sega Computer*, August: 21.

Anonymous. 1988. "Personal Publishing." *Australian Commodore Review Annual*, 64–65.

Anonymous editor. 1986. "Introduction." *Sega Computer*, April: 1. https://ia601708.us.archive.org/13/items/Sega_Computer_1986-04_Nomac_Publishing_NZ/Sega_Computer_1986-04_Nomac_Publishing_NZ.pdf.

Anonymous reviewer. 1983. "The Line between the Toys and Real Business Systems." *Pacific Computer Weekly*, September 5–11, 1983, 8–9.

Anthropy, Anna. 2012. *Rise of the Videogame Zinesters: How Freaks, Normals, Amateurs, Artists, Dreamers, Dropouts, Queers, Housewives, and People Like You Are Taking Back an Art Form.* New York: Seven Stories Press.

Arbesman, Samuel. 2015. "Get under the Hood." *Aeon*, March 2015. https://aeon.co/essays/computers-are-so-easy-that-we-ve-forgotten-how-to-create.

Arrow, Selwyn. 1985. "Personal Computers." In *Looking Back to Tomorrow: Reflections on Twenty-Five Years of Computers in New Zealand*, 105–119. Wellington: New Zealand Computer Society.

Art of Coding. n.d. "Demoscene—The Art of Coding." Accessed June 17, 2020. http://demoscene-the-art-of-coding.net/.

Audit Bureau of Circulations. 1983. "Summary: April 1, 1983 to September 30, 1983." *Audit Bureau of Circulations*, no. 92, 24.

Ault, Roz. 1983. *BASIC Programming for Kids: BASIC Programming on Personal Computers by Apple, Atari, Commodore, Radio Shack, Texas Instruments, Timex Sinclair.* Boston: Houghton Mifflin.

Australian Bureau of Statistics. 1994. *Household Use of Information Technology—February 1994.* ABS Catalogue no. 8128.0. http://www.ausstats.abs.gov.au/ausstats/free.nsf/0/26DDE373E004B1D5CA25722500073A59/$File/81280_0294.pdf.

Australian Bureau of Statistics. 1997. *Household Use of Information Technology—1996.* ABS Catalogue no. 8146.0. http://www.ausstats.abs.gov.au/ausstats/free.nsf/0/0160F55248224B2BCA25722500073A63/$File/81460_1996.pdf.

Australian ICOMOS. 1979. "The Burra Charter: The Australia ICOMOS Charter for Places of Cultural Significance." http://australia.icomos.org/publications/charters/.

Banks, John. 2013. *Co-creating Videogames.* London: Bloomsbury Academic.

Barnes, John G. P. 1982. "The Development of Cultures in Programming Languages." In *Ninth Australian Computer Conference*, edited by A. H. J. Sale and G. Hawthorne, 9–21. Hobart, Tas: Australian Computer Society.

Works Cited

Barr-Hyde, Jeremy. n.d. "Jeremy Barr-Hyde Australian Apple II Software." https://archive.org/details/softwarelibrary_apple_jbh.

Beardon, Colin. 1985. *Computer Culture: The Information Revolution in New Zealand*. Auckland: Reed Methuen.

Benjamin, Walter. 1992a. "The Work of Art in the Age of Mechanical Reproduction." In *Illuminations*, edited by Hannah Arendt, 219–253. London: Fontana.

Benjamin, Walter. 1992b. "Theses on the Philosophy of History." In *Illuminations*, edited by Hannah Arendt, 245–255. London: Fontana.

Beregi, Tamás. 2015. "Hungary." In *Video Games around the World*, edited by Mark J. P. Wolf, 219–234. Cambridge, MA: MIT Press.

Better World Studio. 2015. "Originality in Mobile Games." July. https://betterworldstudio.wordpress.com/2015/07/.

"Birth Control via Your C64." 1988. *Australian Commodore Review* 5 (7): 4.

Bissinger, Buzz. 2015. "Caitlyn Jenner: The Full Story." *Vanity Fair*, July 2015. https://www.vanityfair.com/hollywood/2015/06/caitlyn-jenner-bruce-cover-annie-leibovitz.

Booth, Paul. 2010. *Digital Fandom*. New York: Peter Lang.

Brin, David. 2006. "Why Johnny Can't Code." *Salon*, September 14, 2006. http://www.salon.com/2006/09/14/basic_2/.

Brown, Russell. 2003. "'Blast from Our Past.'" *Unlimited Magazine New Zealand*, September 2003. https://publicaddress.net/hardnews/sportronic-in-beige/.

Brügger, Niels. 2018. *The Archived Web: Doing History in the Digital Age*. Cambridge, MA: MIT Press.

Bruns, Axel. 2008. "FCJ-066 the Future Is User-Led: The Path towards Widespread Produsage." *Fibreculture Journal*, 11. http://eleven.fibreculturejournal.org/fcj-066-the-future-is-user-led-the-path-towards-widespread-produsage/.

Buchsbaum, Walter H., and Robert Mauro. 1979. *Electronic Games: Design, Programming, and Troubleshooting*. New York: McGraw-Hill.

Buckwalter, Len. 1977. *Video Games*. New York: Grosset & Dunlap.

Burnham, Van. 2003. *Supercade: A Visual History of the Videogame Age, 1971–1984*. Cambridge, MA: MIT Press.

Butterfield, Jim. 1978. "Games—Not Just for Fun (Unvarnished Truth about KIM?)." *Creative Computing*, November/December 1978, 104–105.

Campbell-Kelly, Martin. 2003. *From Airline Reservations to Sonic the Hedgehog: A History of the Software Industry*. Cambridge, MA: MIT Press.

Campbell-Kelly, Martin, and William Aspray. 2004. *Computer: A History of the Information Machine*. 2nd ed. Boulder, CO: Westview Press.

Camper, Brett. 2008. "Shareware Games: Between Hobbyist and Professional." In *The Video Game Explosion: A History from PONG to PlayStation and Beyond*, edited by Mark J. P. Wolf, 151–157. Westport, CT: Greenwood Press.

Cass, Stephen. 2014. "The Golden Age of Basic—IEEE Spectrum." *IEEE Spectrum*, May 2014. http://spectrum.ieee.org/tech-talk/computing/software/the-golden-age-of-basic.

Caufield, Anthony, and Nicola Caufield. 2014. *Bedroom to Billions*. London: Gracious Films.

Chalmers, Jackie. 1985. *Getting to Know Your IBM PC*. North Ryde: CCH Australia.

Champion, Erik, ed. 2012a. *Game Mods: Design, Theory and Criticism*. Pittsburgh: ETC Press.

Champion, Erik. 2012b. "Introduction: Mod Mod Glorious Mod." In *Game Mods: Design, Theory and Criticism*, edited by Erik Champion, 11–25. Pittsburgh: ETC Press.

Chan, Rosalie. 2020. "This Electronics Manufacturing Company Changed Its Operations to Make Coronavirus Medical Supplies." *Business Insider*, April 2020. https://www.weforum.org/agenda/2020/04/manufacturing-adafruit-medical-supplies-crisis-relief-covid-coronavirus-pandemic/.

Cifaldi, Frank. 2018. Twitter, August 9, 2018. https://twitter.com/frankcifaldi/status/1027302289600466945?s=20.

Clark, James I. 1985. *A Look inside Video Games*. Milwaukee: Raintree.

Commodore Computer. 1988. "The 3 Best Ways I Know to Introduce Your Family to the World of Computers," featuring John Laws. Advertisement. *Australian Commodore and Amiga Review* 5 (2): 11.

Conley, Tom. 1988. "Translator's Introduction: For a Literary Historiography." In *The Writing of History*, edited by Michel de Certeau, vii–xxiv. New York: Columbia University Press.

Crawford, Geoff. c.1987. "Letter from Poseidon Software to Sega magazine subscribers." Andrew Wheeler Archive.

Deane, John, and Judy Deane. 1980. *Easy Way to Programming in BASIC Using the System 80 Computer*. Sydney: Dick Smith Electronics.

de Certeau, Michel. 1984. *The Practice of Everyday Life*. Berkeley: University of California Press.

de Certeau, Michel. 1997. *The Capture of Speech and Other Political Writings*, edited by Luce Giard and translated by Tom Conley. Minneapolis: University of Minnesota Press.

Works Cited

de Certeau, Michel, and Luce Giard. 1997. "Part III: The Everyday Nature of Communication." In Michel de Certeau, *The Capture of Speech and Other Political Writings*, edited by Luce Giard, 89–139. Minneapolis: University of Minnesota Press.

de Certeau, Michel, and Luce Giard. 1998a. "A Practical Science of the Singular." In Michel de Certeau, Luce Giard, and Pierre Mayol, *The Practice of Everyday Life*, Vol. 2, *Living and Cooking*, translated by Timothy J. Tomasik, 251–256. Minneapolis: University of Minnesota Press.

de Certeau, Michel, and Luce Giard. 1998b. "Ghosts in the City." In Michel de Certeau, Luce Giard, and Pierre Mayol, *The Practice of Everyday Life*, Vol. 2, *Living and Cooking*, translated by Timothy J. Tomasik, 133–143. Minneapolis: University of Minnesota Press.

de Certeau, Michel, Luce Giard, and Pierre Mayol. 1998. *The Practice of Everyday Life*, Vol. 2, *Living and Cooking*, translated by Timothy J. Tomasik. Minneapolis: University of Minnesota Press.

Degiovani, Renato. 2003. "Domestic Pioneers." In *GameBrasilis: Catálogo Jogo Eletrônicos Brasileiros*. Sao Paolo: Senac.

DeMaria, Rusel, and Johnny Lee Wilson. 2002. *High Score! The Illustrated History of Electronic Games*. New York: McGraw-Hill / Osborne.

de Vries, Denise, Angela Ndalianis, Helen Stuckey, and Melanie Swalwell. 2013. Popular Memory Archive. http://www.ourdigitalheritage.org/archive/playitagain/.

Dick Smith Electronics. 1984. "Profit from Your Hobby." Advertisement. *Bits and Bytes*, December 1983/January 1984.

Donovan, Tristan. 2010. *Replay: The History of Video Games*. Lewes, UK: Yellow Ant.

Douglas, Susan J. 1992. "Audio Outlaws: Radio and Phonograph Enthusiasts." In *Possible Dreams: Enthusiasm for Technology in America*, edited by John L. Wright, 45–59. New York: Henry Ford Museum and Greenfield Village.

Douglas, Susan J. 1999. *Listening In: Radio and the American Imagination*. New York: Time Books; Toronto: Random House.

Doyle, Shayne. 1983. "Linking via Amateur Radio." *Bits and Bytes*, December 1983, 86.

Dyer, Sam. 2014. *Commodore 64: A Visual Compendium*. Bath, UK: Bitmap Books.

Edwards, Benj. 2009. "Forty Years of Lunar Lander." *Technologizer*, July 19, 2009. http://www.technologizer.com/2009/07/19/lunar-lander/.

Ensmenger, Nathan. 2012. *The Computer Boys Take Over: Computers, Programmers, and the Politics of Technical Expertise*. Cambridge, MA: MIT Press.

Farrell, Andrew. 1985. "Editorial." *Australian Commodore Review* 2 (3): 2.

Farrell, Andrew. 1987. "Profile: Micro Forte." *Home Computer GEM: Games, Entertainment, Music*, July 1987, 10–14.

Fassone, Riccardo. 2017. "Cammelli and Attack of the Mutant Camels: A Variantology of Italian Video Games of the 1980s." *Well Played Journal* 6 (2): 55–71.

Ferguson, Rik. 2018. Twitter. September 13, 2018. https://twitter.com/rik_ferguson/status/1040006729419943939.

Filatoff, Natalie. 1983. "The Class of '82." *Your Computer: Magazine for Business and Pleasure*, April 1983, 26–31.

Fiske, John. 1989. *Understanding Popular Culture*. London: Routledge.

Foege, Alec. 2013. *The Tinkerers: The Amateurs, DIYers, and Investors Who Make America Great*. New York: Basic Books.

Ford, Lindsay R. 1984. "Lawyer Writes Software." *Online: The Microbee Owner's Journal*, October, 6.

Ford, Lindsay R. 1985. "Software Review: Dreamcards Beats the Odds." *Online: The Microbee Owner's Journal*, February, 21.

Foucault, Michel. 1984. "Nietzsche, Genealogy, History." In *The Foucault Reader*, edited and translated by Paul Rabinow, 76–100. New York: Pantheon.

France, Sharon. 1985. "The Hard Word about Software." *Online: The Microbee Owner's Journal*, December, 24.

Franklin, Seb. 2009. "On Game Art, Circuit Bending and Speedrunning as Counter-Practice: 'Hard' and 'Soft' Nonexistence." *C-Theory* 32 (1–2). http://www.ctheory.net/articles.aspx?id=609.

Franz, Kathleen. 2005. *Tinkering: Consumers Reinvent the Early Automobile*. Philadelphia: University of Pennsylvania Press.

Friedman, Ted. 2005. *Electric Dreams: Computers in American Culture*. New York: New York University Press.

Fuller, Glen. 2012. "V8's 'til '98: The V8 Engine, Australian Nationalism and Automobility." *Global Media Journal—Australian Edition* 6 (1). https://www.hca.westernsydney.edu.au/gmjau/archive/v6_2012_1/glen_fuller_RA.html.

Galloway, Patricia. 2011. "Retrocomputing, Archival Research, and Digital Heritage Preservation: A Computer Museum and ISchool Collaboration." *Library Trends* 59 (4): 623–636.

Gazzard, Alison. 2014. "The Intertextual Arcade: Tracing Histories of Arcade Clones in 1980s Britain." *Reconstruction: Studies in Contemporary Culture* 14 (1).

Gazzard, Alison. 2016. *Now the Chips Are Down: The BBC Micro*. Cambridge, MA: MIT Press.

Works Cited

Giard, Luce. 1997. "Introduction: How Tomorrow Is Already Being Born." In Michel de Certeau, *The Capture of Speech and Other Political Writings*, edited by Luce Giard, vii–xix. Minneapolis: University of Minnesota Press.

Giard, Luce. 1998a. "Introduction to Volume 1: History of a Research Project." In Michel de Certeau, Luce Giard, and Pierre Mayol, *The Practice of Everyday Life*, Vol. 2, *Living and Cooking*, translated by Timothy J. Tomasik, xiii–xxxiii. Minneapolis: University of Minnesota Press.

Giard, Luce. 1998b. "Times and Places." In Michel de Certeau, Luce Giard, and Pierre Mayol, *The Practice of Everyday Life*, Vol. 2, *Living and Cooking*, translated by Timothy J. Tomasik, xxxv–xlv. Minneapolis: University of Minnesota Press.

Gielens, Jaro. 2000. *Electronic Plastic*. Berlin: Gestalten Verlag.

Gillespie, Tarlton. 2009. *Wired Shut: Copyright and the Shape of Digital Culture*. Cambridge, MA: MIT Press.

Ginzburg, Carlo. 1992. *The Cheese and the Worms: The Cosmos of a Sixteenth-Century Miller*, translated by John and Anne Tedeschi. New York: Penguin.

Graham, Ian. 1982. *Usborne Guide to Computer and Video Games*. London: Usborne.

Gruber Garvey, Ellen. 2003. "Scissorizing and Scrapbooks: Nineteenth-Century Reading, Remaking, and Recirculating." In *New Media: 1740–1915*, edited by Lisa Gitelman and Geoffrey Pingree, 207–227. Cambridge, MA: MIT Press.

Guins, Raiford. 2014. *Game After: A Cultural Study of Video Game Afterlife*. Cambridge, MA: MIT Press.

Gunness, Jacob. 1999. "A Chat with Dorothy Millard, Australian Adventure Author." Solution Archive. http://solutionarchive.com/interview_dorothy/.

Gunning, Tom. 1989. "An Aesthetic of Astonishment." *Art & Text* 34:31–45.

Gunning, Tom. 2003. "Re-newing Old Technologies: Astonishment, Second Nature, and the Uncanny in Technology from the Previous Turn-of-the-Century." In *Rethinking Media Change: The Aesthetics of Transition*, edited by David Thorburn and Henry Jenkins, 39–60. Cambridge, MA: MIT Press.

HackerRank. 2018. "2018 Developer Skills Report." http://research.hackerrank.com/developer-skills/2018/.

"HackFest | KansasFest." n.d. Accessed October 31, 2017. https://www.kansasfest.org/hackfest/.

Haddon, Leslie. 1988. "The Home Computer, the Making of a Consumer Electronic." *Science as Culture* 2:7–51.

Hague, James. 2002. *Halcyon Days: Interview with Classic Computer and Video Game Programmers*. http://www.dadgum.com/halcyon/.

Haigh, Thomas. 2011. "The History of Information Technology." *Annual Review of Information Science and Technology* 45 (1): 431–487.

Halegoua, Germaine R. 2016. "Introduction: Locating Emerging Media." In *Locating Emerging Media*, edited by Germaine R. Halegoua and Ben Aslinger, 1–13. New York: Routledge.

Hall, Matthew. 2013. "Statement about 'Jewels of Sancara Island.'" Popular Memory Archive. http://www.ourdigitalheritage.org/archive/playitagain/games/jewels-of-sancara-island/.

Hansen, Miriam Bratu. 1999. "Benjamin and Cinema: Not a One-Way Street." *Critical Inquiry* 25 (2): 306–343.

Hansen, Nicolai Brodersen, Rikke Toft Nørgård, and Kim Halskov. 2014. "Crafting Code at the Demo-Scene." In *Conference on Designing Interactive Systems*, edited by Ron Wakkary and Steve Harrison, 35–38. Vancouver: Association for Computing Machinery.

Haring, Kristen. 2007. *Ham Radio's Technical Culture*. Cambridge, MA: MIT Press.

Harrigan, Pat, and Matthew Kirschenbaum, eds. 2016. *Zones of Control: Perspectives on Wargaming*. Cambridge, MA: MIT Press.

Harrison, R. 1985. "History of Hacking in Horstralia, Part 1." *Online: The Microbee Owner's Journal*, June, 8–10.

Harrison, Roger, ed. 1985a. *Electronics Today's Circuits Cookbook #5*. Sydney: Eastern Suburbs Newspapers.

Harrison, Roger, ed. 1985b. *Microbee Hacker's Handbook: Hard and Soft Projects for Bees of All Vintages*. Sydney: Federal Publishing.

Hearn, Louisa. 2006. "How Rudie Brought Apple to Australia." *Sydney Morning Herald*, March 30, 2006. www.smh.com.au/news/breaking/how-rudie-brought-apple-to-australia/2006/03/30/1143441250392.html.

Heilborn, John. 1988. *Commodore 128 Troubleshooting and Repair*. Indianapolis: Howard W. Sams / Macmillan.

Henn, Steve. 2014. "When Women Stopped Coding." *Planet Money*, NPR, October 21, 2014. http://www.npr.org/sections/money/2014/10/21/357629765/when-women-stopped-coding.

Herz, J. C. 1997. *Joystick Nation: How Videogames Gobbled Our Money, Won Our Hearts and Rewired Our Minds*. London: Abacus.

Hicks, Mar. 2017. *Programmed Inequality: How Britain Discarded Women Technologists and Lost Its Edge in Computing*. Cambridge, MA: MIT Press.

Hilbert, Ernest. 2004. "Flying Off the Screen: Observations from the Golden Age of the American Video Game Arcade." In *Gamers: Writers, Artists, and Programmers on the Pleasures of Pixels*, edited by Shanna Compton, 57–69. New York: Soft Skull.

Hills, Matt. 2013. "Fiske's 'Textual Productivity' and Digital Fandom: Web 2.0 Democratization versus Fan Distinction." *Participations: Journal of Audience and Reception Studies* 10 (1). http://www.participations.org/Volume 10/Issue 1/9 Hills 10.1.pdf.

Hoess, Rudi. 1978. "The Games People Play." *Sydney Morning Herald*, July 25, 1978.

Huang, Andrew. 2003. *Hacking the Xbox: An Introduction to Reverse Engineering*. San Francisco: No Starch Press.

Hughes, Matthew. 2018. "Report: 80's Kids Started Programming at an Earlier Age than Today's Millennials." *The Next Web*, January 23, 2018. https://thenextweb.com/dd/2018/01/23/report-80s-kids-started-programming-at-an-earlier-age-than-todays-millennials/?utm_source=facebook.com&utm_medium=referral&utm_content=Report%3A+80%27s+kids+started+programming+at+an+earlier+age+than+today%27s+millenni.

Huhtamo, Erkki. 2005. "Slots of Fun, Slots of Trouble: An Archaeology of Arcade Gaming." In *Handbook of Computer Game Studies*, edited by Joost Raessens and Jeffrey H. Goldstein, 3–21. Cambridge, MA: MIT Press.

Huhtamo, Erkki. 2011. "Dismantling the Fairy Engine: Media Archaeology as Topos Study." In *Media Archaeology: Approaches, Applications, and Implications*, edited by Erkki Huhtamo and Jussi Parikka, 27–47. Berkeley: University of California Press.

Huhtamo, Erkki, and Jussi Parikka. 2011. "Introduction: An Archaeology of Media Archaeology." In *Media Archaeology: Approaches, Applications, and Implications*, edited by Erkki Huhtamo and Jussi Parikka, 1–21. Berkeley: University of California Press.

Inman, Don, Ramon Zamora, and Bob Albrecht. 1981. *More TRS-80 BASIC*. New York: Wiley.

Ippolito, Jon, and Richard Rinehart. 2014. *Re-collection Art, New Media, and Social Memory*. Cambridge, MA: MIT Press.

Ironmonger, D. S., C. W. Lloyd-Smith, and F. Soupourmas. 2000. "New Products of the 1980s and 1990s: The Diffusion of Household Technology in the Decade 1985–1995." *Prometheus* 18 (4): 403–415. https://doi.org/10.1080/0810902002000851.

Jakić, Bruno. 2014. "Galaxy and the New Wave: Yugoslav Computer Culture in the 1980s." In *Hacking Europe: From Computer Cultures to Demoscenes*, edited by Gerard Alberts and Ruth Oldenziel, 129–150. London: Springer.

Jenkins, Henry. 1992. *Textual Poachers: Television Fans and Participatory Culture*. New York: Routledge.

Jenkins, Henry. 2006. *Fans, Bloggers, and Gamers: Exploring Participatory Culture*. New York: New York University Press.

Jenkins, Henry. 2007. "Afterword: The Future of Fandom." In *Fandom: Identities and Communities in a Mediated World*, edited by Jonathan Gray, Cornel Sandvoss, and C. Lee Harrington, 357–364. New York: New York University Press.

Jenkins, Henry. 2010. "Introduction: Why Fiske Still Matters." In *Television Culture*, by John Fiske, 2nd ed., xv–xli. London: Routledge.

Jones, Stephen. 2011. *Synthetics: Aspects of Art and Technology in Australia*. Cambridge, MA: MIT Press.

Jones, Steven E. 2008. *The Meaning of Video Games: Gaming and Textual Strategies*. New York: Routledge.

Kent, Steven L. 2001. *The Ultimate History of Video Games: From Pong to Pokemon and beyond—The Story behind the Craze That Touched Our Lives and Changed the World*. New York: Three Rivers Press.

Kenyon, Philip. n.d. "Sega Users Club." Unpublished letter. Aaron Wheeler Archive.

Kenyon, Philip. 1984. "Welcome to the Sega Users Club." *Sega Computer*, August 1984, 1.

Keogh, Brendan. 2017. "Who Else Makes Videogames? Considering Informal Development Practices." In *Digital Games Research Association Conference*. Melbourne. http://digra2017.com/static/Extended Abstracts/146_DIGRA2017_EA_Keogh_Informal_Development.pdf.

Keogh, Brendan. 2019. "From Aggressively Formalised to Intensely In/Formalised: Accounting for a Wider Range of Videogame Development Practices." *Creative Industries Journal* 12 (1): 14–33.

King, Brad, and John Borland. 2003. *Dungeons and Dreamers: The Rise of Computer Game Culture from Geek to Chic*. New York: McGraw-Hill / Osborne.

Kirkpatrick, Graeme. 2007. "Meritums, Spectrums and Narrative Memories of 'Pre-Virtual' Computing in Cold War Europe 1." *Sociological Review* 55 (2): 227–250.

Kirkpatrick, Graeme. 2012. "Game Studies—Constitutive Tensions of Gaming's Field: UK Gaming Magazines and the Formation of Gaming Culture 1981–1995." *Game Studies* 12 (1). http://gamestudies.org/1201/articles/kirkpatrick.

Kirkpatrick, Graeme. 2014. "Making Games Normal: Computer Gaming Discourse in the 1980s." *New Media and Society* 18 (8): 1439–1454.

Kirkpatrick, Graeme. 2015. *The Formation of Gaming Culture: UK Gaming Magazines, 1981–1995*. Houndsmills, UK: Palgrave Macmillan.

Kirkpatrick, Graeme. 2017. "Early Games Production, Gamer Subjectivation and the Containment of the Ludic Imagination." In *Fans and Videogames: Histories, Fandom, Archives*, edited by Melanie Swalwell, Helen Stuckey, and Angela Ndalianis, 19–37. New York: Routledge.

Kitch, Bob. n.d. "History of VZ Computers!" VZ-Alive. Accessed June 26, 2019. http://vzalive.bluebilby.com/history/.

Kline, Stephen, Nick Dyer-Witheford, and Greig De Peuter. 2003. *Digital Play: The Interaction of Technology, Culture and Marketing*. Montreal: McGill-Queen's University Press.

Works Cited

Kocurek, Carly. 2015. *Coin-Operated Americans: Rebooting Boyhood at the Video Game Arcade*. Minneapolis: University of Minnesota Press.

Kong Tin, Harvey A. 1986. "Hot Copter Ie. Laser Hawk Development Materials." https://archive.org/details/HotCopter.

Kowalski, John. 1996. "GLOOM 3-D V1.2 DEMONSTRATION GRAPHICS ENGINE." http://users.axess.com/twilight/sock/gloom/gloom.asm.

Kubey, Craig. 1982. *The Winners' Book of Video Games*. London: W. H. Allen.

Lally, Elaine. 2002. *At Home with Computers*. Oxford: Berg.

Laughton, Alan. 2013. "Comment on Hoards of the Deep Realm." Popular Memory Archive. http://www.ourdigitalheritage.org/archive/playitagain/games/hoards-of-the-deep-realm/.

Lean, Tom. 2016. *Electronic Dreams: How 1980s Britain Learned to Love the Computer*. London: Bloomsbury.

Lee, Frank. 1985. "Computing on the Cheap." *Your Computer: Magazine for Business and Pleasure*, August 1985: 19–30.

"Letters to the Editor." 1984. *Online: The Microbee Owner's Journal*, October, 3.

Levi, Giovanni. n.d. "Biography and Microhistory (Abstract)." University of Valencia. http://www.uv.es/retpb/docs/Florencia/Giovanni Levi.pdf.

Levi, Giovanni. 2001. "On Microhistory." In *New Perspectives on Historical Writing*, edited by Peter Burke, 2nd ed., 97–119. University Park: Pennsylvania State University.

Lialina, Olia, and Dragan Espenschied, eds. 2009. *Digital Folklore: To Computer Users, with Love and Respect*. Stuttgart: Merz und Solitude.

Lindsay, Christina. 2003. "From the Shadows: Users as Designers, Producers, Marketers, Distributors, and Technical Support." In *How Users Matter: The Co-construction of Users and Technology*, edited by Trevor J. Pinch and Nelly Oudshoorn, 29–50. Cambridge, MA: MIT Press.

Lindsay, E. 1985. "Heard around the Traps." *Online: The Microbee Owner's Journal*, August, 40.

Lindsay, Eric. 1983. "Designing Your Own Projects." *Your Computer: Magazine for Business and Pleasure* 3 (5): 114–115.

Lindsay, Eric, ed. 1989. *Games #1: Quick Reference (Applix 1616 Shareware Manual)*. Faulconbridge: Eric Lindsay. http://www.ericlindsay.com/applix/swgames1.pdf.

Lindsay, Eric, and Tom Moffat. 1982. "Two Views of the Microbee." *Electronics Today International*, December 1982, 94–97.

Lloyd, Genevieve. 1984. *Man of Reason: "Male" and "Female" in Western Philosophy*. London: Methuen.

Lobato, Ramon. 2012. *Shadow Economies of Cinema: Mapping Informal Film Distribution*. London: BFI / Palgrave Macmillan.

Loguidice, Bill, and Matt Barton. 2009. *Vintage Games: An Insider Look at the History of Grand Theft Auto, Super Mario, and the Most Influential Games of All Time*. Burlington, MA: Focal Press / Elsevier.

Longhurst, Brian. 2007. *Cultural Change and Ordinary Life*. Maidenhead, UK: McGraw-Hill Education.

Love, Darren, and Guy Hancock. 1984. *Astounding Arcade Games for the John Sands Sega*. London: Interface.

Lowood, Henry. 2001. "The Hard Work of Software History." *RBM: A Journal of Rare Books, Manuscripts, and Cultural Heritage* 2 (2): 141–160. http://www.ala.org/ala/acrl/acrlpubs/rbm/backissuesvol2no/lowood.PDF.

Lowood, Henry. 2016. "It Is What It Is, Not What It Was." *Refractory: A Journal of Entertainment Media* 27. https://refractory-journal.com/henry-lowood/.

Lowood, Henry, and Raiford Guins, eds. 2016. *Debugging Game History: A Critical Lexicon*. Cambridge, MA: MIT Press.

"Machine Code Made Easy." n.d. Auckland City Library Ephemera: Technology, 1980s.

Maher, Jimmy. 2012. *The Future Was Here: The Commodore Amiga*. Cambridge, MA: MIT Press.

Mailland, Julien, and Kevin Driscoll. 2017. *Minitel: Welcome to the Internet*. Cambridge, MA: MIT Press.

Marentes, Nickolas. n.d.a. "Donut Dilemma." http://nickmarentes.com/ProjectArchive/donut.html.

Marentes, Nickolas. n.d.b. "Donut Dilemma (CoCo)." http://nickmarentes.com/ProjectArchive/donutcc.html.

Marentes, Nickolas. n.d.c. "Space Intruders." http://nickmarentes.com/ProjectArchive/intruders.html.

Marentes, Nickolas. n.d.d. "Stellar Odyssey." http://nickmarentes.com/ProjectArchive/stellar1.html.

Marshall, P. David. 2004. *New Media Cultures*. London: Hodder.

Maynard, F. K. 1985. "Letter." *Sega Computer*, January/February 1985: 2.

McBeth. 1988. "'Penny Stock' Coup by Computer Whiz." *Business Review Weekly*, August 12, 1988.

McCracken, Harry. 2014. "Fifty Years of BASIC, the Programming Language That Made Computers Personal." *Time*, April 2014. http://time.com/69316/basic/.

McGinley, Jim. 2012. "Inspiration from the Trash: The TRS-80's Lost Game Designs." Game Developers Conference. http://www.gdcvault.com/play/1015922/Inspiration-from-the-Trash-The.

Meggs, Philip. 2016. *Meggs' History of Graphic Design*. 6th ed. Hoboken, NJ: Wiley.

Mendham, Trevor. 1986. "All Work and No Play in the Office Is Bad for Business." *Sydney Morning Herald*, February 24, 1986, 14.

Micchia, Shayne. 1985. *The Bewildered Parent's Guide to Computer Programming*. Melbourne: Pitman Press.

Monro, Don. 1982. *Learning BASIC on the VIC 20*. Sydney: Prentice-Hall.

Montfort, Nick, and Ian Bogost. 2009. *Racing the Beam: The Atari Video Computer System*. Cambridge, MA: MIT Press.

Morley, David. 2003. "The Nationwide Audience." In *The Audience Studies Reader*, edited by Will Brooker and Deborah Jermyn, 91–104. London: Routledge.

Morris, Sue. 1999. "Online Gaming Culture: An Examination of Emerging Forms of Production and Participation in Multiplayer First-Person-Shooter Gaming." Game-Culture. http://www.game-culture.com/articles/onlinegaming.html.

Morton, Andrew, and Paul Berger. 1986. "16-Bit Computer." *Electronics Today International*, December 1986, 54–57.

Moss, Richard. 2018. *The Secret History of Mac Gaming*. London: Unbound.

Neumark, Norie. 1993. "Diagnosing the Computer User: Addicted, Infected or Technophiliac?" *Media Information Australia*, no. 69, 80–87.

Newman, James. 2012. *Best Before: Videogames, Supersession and Obsolescence*. London: Routledge.

Newman, Michael Z. 2017. *Atari Age: The Emergence of Video Games in America*. Cambridge, MA: MIT Press.

Nylund, Niklas. 2018. "Constructing Digital Game Exhibitions: Objects, Experiences, and Context." *Arts* 7 (4). https://doi.org/10.3390/arts7040103.

O'Hara, Rob. 2006. *Commodork: Sordid Tales from a BBS Junkie*. Yukon, OK: OHara Press.

Oliver, Ron. 1984. *Get Started on Your Micro*. Carlton, Australia: Pitman Press.

Orland, Kyle. 2019. "Nintendo Issues DMCA Takedown for Super Mario Bros. Commodore 64 Port." *Ars Technica*, April 24, 2019. https://arstechnica.com/gaming/2019/04/nintendo-issues-dmca-takedown-for-super-mario-bros-commodore-64-port/.

Oudshoorn, Nelly, and Trevor Pinch, eds. 2003. *How Users Matter: The Co-construction of Users and Technology*. Cambridge, MA: MIT Press.

Parikka, Jussi. 2010. "Archaeologies of Media Art: Jussi Parikka in Conversation with Garnet Hertz." *C-Theory.* https://journals.uvic.ca/index.php/ctheory/article/view/14750/5621.

Parikka, Jussi. 2012. *What Is Media Archaeology?* Cambridge: Polity.

Paull, John, and Adrian Kovac. 1984. *Introductory Computer Course: Student Notes.* Bondi: Computer Seminars (Australia).

Philipson, Graeme. 2003. "In the Hype Cycle, Failure Always Precedes Success." *The Age*, April 8, 2003. http://www.theage.com.au/articles/2003/04/07/1049567604685.html.

Philipson, Graeme. 2004. "Time Machine: A Brief History of Computing." Core Memory. http://www.thecorememory.com/TimeMachine.pdf.

Poster, Mark. 1992. "The Question of Agency: Michel de Certeau and the History of Consumerism." *Diacritics* 22 (2): 94–107.

Powell, Gareth. 1985. "Manipulating Images: A New Role for the Computer." *Australian Apple Review* 2 (6): 19–22.

Powell, Gareth. 1986. "Telecomputing, Part 1." *Australian Apple Review* 3 (9): 46–48.

Purdy, Kevin. 2020. "How iFixit Built Its Free Medical Database." *iFixit.* https://www.ifixit.com/News/41634/how-ifixit-built-its-free-medical-database.

Rankin, Joy Lisi. 2018. *A People's History of Computing in the United States.* Cambridge, MA: Harvard University Press.

Rechert, Klaus, Patricia Falcao, and Tom Ensom. 2016. "Introduction to an Emulation-Based Preservation Strategy for Software-Based Artworks." Tate. http://www.tate.org.uk/download/file/fid/105887.

Rehak, Bob. 2008. "The Rise of the Home Computer." In *The Video Game Explosion: A History from PONG to PlayStation and Beyond*, edited by Mark J. P. Wolf, 75–80. Westport, CT: Greenwood Press.

Reunanen, Markku, and Antti Silvast. 2009. "Demoscene Platforms: A Case Study on the Adoption of Home Computers." In *IFIP Conference on History of Nordic Computing*, edited by J. Imagliazzo, T. Järvi, and P. Paju, 289–301. Berlin: Springer.

Richardson, R. 1985a. "Realities of the Workplace: The Case History of a Small Business Starting to Use a Commodore 64." *Australian Commodore Review* 2 (8): 36–37.

Richardson, R. 1985b. "The First Business Software." *Australian Commodore Review* 2 (9): 45–46.

Richardson, R. 1986. "Using the Commodore 64 and SID in the Office." *Australian Commodore Review* 3 (1): 33–35.

Richardson, Ric. 1985. "The Next Home Entertainment Revolution." *Australian Apple Review* 2 (8): 26–27.

Rieger, Oya Y., Tim Murray, Madeleine Casad, Desiree Alexander, Dianne Dietrich, Jason Kovari, Liz Muller, Michelle Paolillo, and Danielle K. Mericle. 2015. "Preserving and Emulating Digital Art Objects." https://ecommons.cornell.edu/handle/1813/41368.

Rosen, Jay. n.d. "The People Formerly Known as the Audience." *Huffington Post*. Accessed July 16, 2013. http://www.huffingtonpost.com/jay-rosen/the-people-formerly-known_1_b_24113.html.

Rosenthal, David S. H. 2015. "Emulation and Virtualization as Preservation Strategies." Andrew W. Mellon Foundation. https://mellon.org/media/filer_public/0c/3e/0c3eee7d-4166-4ba6-a767-6b42e6a1c2a7/rosenthal-emulation-2015.pdf.

ross. 2014. "Comment on Donut Dilemma." Popular Memory Archive. http://www.ourdigitalheritage.org/archive/playitagain/games/donut-dilemma/.

Rowe, Jamieson. 1974. "Build Your Own Digital Computer! (Part 1 of a Series of Articles)." *Electronics Australia*, August 1974: 42–47.

Rowlands, Grant. 1978. "It's an Era of Friendly Little Computers. . . ." *Sydney Morning Herald*, January 9, 1978, 8.

"Rupert Rhthym." 1989. *The Rainbow*, January 1989.

Saarikoski, Petri, and Jaakko Suominen. 2009. "Computer Hobbyists and the Gaming Industry in Finland." *IEEE Annals of the History of Computing* 31 (3): 20–33. https://doi.org/10.1109/MAHC.2009.39.

Saarikoski, Petri, Jaakko Suominen, and Markku Reunanen. 2017. "Pac-Man for the Vic-20." *Well Played Journal* 6 (2): 7–31.

Savetz, Kevin. 2012. *Terrible Nerd: True Tales of Growing up Geek!* Portland: Savetz Publishing.

Schleiner, Anne-Marie. 2002. "2 Reviews: Untitled Game and Ego Image Shooter (Review)." Nettime. http://amsterdam.nettime.org/Lists-Archives/nettime-l-0203/msg00061.html.

Schlombs, Corinna. 2006. "Toward International Computing History." *IEEE Annals of the History of Computing* 28 (1): 107–108.

Schmitt, John, and Jonathan Wadsworth. 2002. *Give PC's a Chance: Personal Computer Ownership and the Digital Divide in the United States and Great Britain*. London: London School of Economics and Political Science.

Schulz, Celine, and Stefan Wagner. 2008. "Outlaw Community Innovations." *International Journal of Innovation Management* 12 (3): 399–418.

Scorgie, Michael, and Anne Magnus. 1984. *Accounting on Your IBM-PC*. Sydney: Prentice-Hall of Australia; Reston, VA: Reston Publishing.

Scott, Jason. 2005. *BBS: The Documentary*. Waltham, MA: Bovine Ignition Systems.

Scott, Jason. 2019. Twitter, April 15, 2019. https://twitter.com/textfiles/status/1117475339536609280.

Seremetakis, C. Nadia. 1996. "The Memory of the Senses, Part I: Marks of the Transitory." In *The Senses Still: Perception and Memory as Material Culture in Modernity*, 1–18. Chicago: University of Chicago Press.

Shaw, Adrienne. 2019. "Archival Serendipity, Excitement, and Whiplash: Affects and Collisions in Doing LGBTQ Game History." *ROM Chip: A Journal of Game Histories* 1 (2). https://romchip.org/index.php/romchip-journal/article/view/92.

Silicon Chip Publications. n.d. "About Silicon Chip." *Silicon Chip Magazine*. Accessed June 18, 2019. www.siliconchip.com.au/Help/About.

Silverstone, Roger. 1994. *Television and Everyday Life*. London: Routledge.

Simon, Bart. 2007. "Geek Chic: Machine Aesthetics, Digital Gaming, and the Cultural Politics of the Case Mod." *Games and Culture* 2 (3): 175–193.

"SIRIUS 7 (AMIGA—FULL GAME)—YouTube." 2013. Zeusdaz—The Unemulated Retro Game Channel. https://www.youtube.com/watch?v=G6IZt2ZmldA.

Smith, Bob. n.d.a. "Ant Attack." Accessed June 19, 2019. https://bobs-stuff.itch.io/ant-attack.

Smith, Bob. n.d.b. "CroZXy Road." Accessed June 19, 2019. https://bobs-stuff.itch.io/crozxy-road.

Smith, Bob. n.d.c. "Noir Shapes." Accessed June 19, 2019. https://bobs-stuff.itch.io/noir-shapes.

Smith, Bob. n.d.d. "Welcome to Bob's Stuff." Accessed June 19, 2019. https://www.bobs-stuff.co.uk/.

Smith, Bob. n.d.e. "ZXagon." Accessed June 19, 2019. https://bobs-stuff.itch.io/zxagon.

"The Software Library: Apple Computer." 2019. https://archive.org/details/softwarelibrary_apple?%2F=&sort=-downloads&page=2.

Sotamaa, Olli. 2009. "The Player's Game: Towards Understanding Player Production among Computer Game Cultures." PhD thesis, University of Tampere, Finland.

Stebbins, Robert A. 2001. "Serious Leisure." *Society*, May/June, 53–57.

Sterling, Bruce. n.d. "The Dead Media Project: A Modest Proposal and a Public Appeal." http://www.deadmedia.org/modest-proposal.html.

Street, C. A. 1983. *Information Handling for the ZX Spectrum*. Maidenhead, UK: McGraw-Hill.

Stuckey, Helen. 2013. "Australian Pioneers: Melbourne House and the Tyranny of Distance." Popular Memory Archive. http://www.ourdigitalheritage.org/archive/playitagain/australian-pioneers-melbourne-house-and-the-tyranny-of-distance-3/.

Works Cited

Stuckey, Helen. 2014. "Exhibiting The Hobbit: A Tale of Memories and Microcomputers." *Kinephanos*. http://www.kinephanos.ca/2014/the-hobbit/.

Stuckey, Helen, and Melanie Swalwell. 2014. "Retro-computing Community Sites and the Museum." In *Handbook of Digital Games*, edited by Mario C. Angelides and Harry Agius, 523–547. Hoboken, NJ: IEEE / Wiley.

Stuckey, Helen, Melanie Swalwell, Angela Ndalianis, and Denise de Vries. 2015. "Remembering & Exhibiting Games Past: The Popular Memory Archive." *Transactions of DiGRA* 2 (1). http://todigra.org/index.php/todigra/article/view/40.

Sumner, James. 2012. "'Today, Computers Should Interest Everybody': The Meanings of Microcomputers." *Studies in Contemporary History* 9: 307–315.

Sun Herald. 1985. "Fun Computer Books." *Sun Herald*, Sydney, March 3, 1985.

Švelch, Jaroslav. 2013. "Indiana Jones Fights the Communist Police: Local Appropriation of the Text Adventure Genre in the 1980s Czechoslovakia." In *Gaming Globally: Production, Play, and Place*, edited by Nina B. Huntemann and Ben Aslinger, 163–181. New York: Palgrave Macmillan.

Švelch, Jaroslav. 2017. "Keeping the Spectrum Alive: Platform Fandom in a Time of Transition." In *Fans and Videogames: Histories, Fandom, Archives*, edited by Melanie Swalwell, Helen Stuckey, and Angela Ndalianis, 57–74. New York: Routledge.

Švelch, Jaroslav. 2018. *Gaming the Iron Curtain: How Teenagers and Amateurs in Communist Czechoslovakia Claimed the Medium of Computer Games*. Cambridge, MA: MIT Press.

Swaine, Michael, and Paul Freiberger. 2000. *Fire in the Valley: The Making of the Personal Computer*. 2nd ed. New York: McGraw-Hill.

Swalwell, Melanie. 2002. "Aesthetics and Hyper/Aesthetics: Rethinking the Senses in Contemporary Media Contexts." Unpublished PhD thesis, University of Technology, Sydney. http://hdl.handle.net/2100/386.

Swalwell, Melanie. 2007. "The Remembering and the Forgetting of Early Digital Games: From Novelty to Detritus and Back Again." *Journal of Visual Culture* 6 (2): 255–273. https://doi.org/10.1177/1470412907078568.

Swalwell, Melanie. 2008a. "1980s Home Coding: The Art of Amateur Programming." In *Aotearoa Digital Arts Reader*, edited by Stella Brennan and Su Ballard, 193–201. Auckland: Clouds / ADA.

Swalwell, Melanie. 2008b. "Kinaesthetic Responsiveness: A Neglected Pleasure." In *The Pleasures of Computer Games: Essays on Cultural History, Theory and Aesthetics*, edited by Melanie Swalwell and Jason Wilson, 72–93. Jefferson, NC: McFarland.

Swalwell, Melanie. 2009. "Towards the Preservation of Local Computer Game Software: Challenges, Strategies, Reflections." *Convergence: The International Journal of Research into New Media Technologies* 15 (3): 263–279. https://doi.org/10.1177/1354856509105107.

Swalwell, Melanie. 2010. "Hobbyist Computing in 1980s New Zealand: Games and the Popular Reception of Microcomputers." In *Return to Tomorrow: 50 Years of Computing in New Zealand*, edited by Janet Toland, 157–169. Wellington: New Zealand Computing Society.

Swalwell, Melanie. 2015. "New Zealand." In *Video Games around the World*, edited by Mark J. P. Wolf, 377–391. Cambridge, MA: MIT Press.

Swalwell, Melanie. 2016. "Classic Gaming." In *Debugging Game History: A Critical Lexicon*, edited by Henry Lowood and Raiford Guins, 45–52. Cambridge, MA: MIT Press.

Swalwell, Melanie. 2017a. "Moving on from the Original Experience: Philosophies of Preservation and Dis/play in Game History." In *Fans and Videogames: Histories, Fandom, Archives*, edited by Melanie Swalwell, Helen Stuckey, and Angela Ndalianis, 213–233. New York: Routledge.

Swalwell, Melanie. 2017b. Twitter, July 13, 2017. https://twitter.com/melswal/status/885337238522839041.

Swalwell, Melanie, ed. 2021. *Game History and the Local*. London: Palgrave.

Swalwell, Melanie, and Janet Bayly. 2010. "More than a Craze: Photographs of New Zealand's Early Digital Games Scene (Exhibition)." Mahara Gallery. http://www.maharagallery.org.nz/MoreThanACraze/index.php.

Swalwell, Melanie, and Michael Davidson. 2016. "Game History and the Case of 'Malzak': Theorising the Manufacture of 'Local Product' in 1980s New Zealand." In *Locating Emerging Media*, edited by Benjamin Aslinger and Germaine Halegoua, 85–105. New York: Routledge.

Swalwell, Melanie, and Erik Loyer. 2006. "Cast-offs from the Golden Age." *Vectors: Journal of Culture and Technology in a Dynamic Vernacular* 3. http://vectorsjournal.org/projects/index.php?project=66.

Swalwell, Melanie, Helen Stuckey, and Angela Ndalianis, eds. 2017. *Fans and Videogames: Histories, Fandom, Archives*. New York: Routledge.

Swalwell, Melanie, and Denise de Vries. 2013. "Collecting and Conserving Code: Challenges and Strategies." *Scan: Journal of Media Arts Culture* 10 (2). http://scan.net.au/scn/journal/vol10number2/Melanie-Swalwell.html.

Symons, Ross. 1984. *First Steps in Machine Code on the Commodore 64*. London: Corgi.

Symons, Ross. 1985. *Your 64 Megabasic: Extended BASIC for the Commodore 64*. London: Interface.

Takahashi, Y. 2000. "A Network of Tinkerers: The Advent of the Radio and Television Receiver Industry in Japan." *Technology and Culture* 41 (3): 460–484.

Tanton, Robert. n.d. "Super Dick's Super 80." *Your Computer: Magazine for Business and Pleasure* 2 (7): 62–63.

Taylor, T. L. 2006. *Play between Worlds*. Cambridge, MA: MIT Press.

TestSheepNZ. 2014. "Programming—It Was Acceptable in the 80s...." http://testsheepnz.blogspot.co.nz/2014/02/programming-it-was-acceptable-in-80s.html.

Thomson, Mark. 2002a. *Rare Trades: Making Things by Hand in the Digital Age*. Sydney: Harper Collins.

Thomson, Mark. 2002b. *The Complete Blokes and Sheds*. Sydney: Harper Collins.

Thomson, Mark. 2007. *Blokes and Sheds: Makers, Breakers and Fixers*. Sydney: Harper Collins.

Tinn, Honghong. 2011. "From DIY Computers to Illegal Copies: The Controversy over Tinkering with Microcomputers in Taiwan, 1980–1984." *IEEE Annals of the History of Computing* 33 (2): 75–88.

Tomasik, Timothy J. 2001. "Certeau á La Carte: Translating Discursive Terroir in The Practice of Everyday Life: Living and Cooking." *South Atlantic Quarterly* 100 (2): 519–542.

Turner, Fred. 2006. *From Counterculture to Cyberculture*. Chicago: University of Chicago Press.

van Dijck, Jose. 2009. "Users Like You? Theorizing Agency in User-Generated Content." *Media, Culture and Society* 31 (1): 41–58.

Veraart, F. 2011. "Losing Meanings: Computer Games in Dutch Domestic Use, 1975–2000." *IEEE Annals of the History of Computing* 33 (1): 52–65.

Veraart, F. 2014. "Transnational (Dis)connection in Localizing Personal Computing in the Netherlands, 1975–1990." In *Hacking Europe: From Computer Cultures to Demoscenes*, edited by Gerard Alberts and Ruth Oldenziel, 25–48. London: Springer.

Vignau, Antoine. n.d. "Index of Projects/X." http://www.brutaldeluxe.fr/projects/x/.

Vignau, Antoine. 2017. Twitter, September 14, 2017. https://twitter.com/antoine_vignau/status/908010980168355841.

Villordsutch. 2016. "Crozxy Road (Crossy Road)." YouTube. https://www.youtube.com/watch?v=KmSicIYkCQw.

Wade, Alex. 2016. *Playback: A Genealogy of 1980s British Videogames*. New York: Bloomsbury.

Warren, Jim, Jr., ed. 1977. *First West Coast Computer Faire Conference Proceedings*. San Francisco: Computer Faire, inside cover.

Warren, Jim, Jr., ed. 1978. *Second West Coast Computer Faire Conference Proceedings*. San Jose: Computer Faire, inside cover.

Wasiak, Patryk. 2014. "Playing and Copying: Social Practices of Home Computer Users in Poland during the 1980s." In *Hacking Europe: From Computer Cultures* to *Demoscenes*, edited by Gerard Alberts and Ruth Oldenziel, 129–150. London: Springer.

West, Garry. 1987. "Microbee Turns to Price for Aid." *Australian Financial Review*, September 18, 1987.

Wheeler, Aaron. 2008a. "An Interview with Dean Hodgson." Sega SC-3000 Survivors. http://www.sc-3000.com/images/stories/interviews/DeanHodgson/InterviewDeanHodgson.pdf.

Wheeler, Aaron. 2008b. "An Interview with Robert B. Brian." Sega SC-3000 Survivors. http://www.sc-3000.com/ (site discontinued).

Wheeler, Aaron, and Michael Davidson. 2008. "SC3K Tape Software List." http://homepages.ihug.co.nz/~atari/SC3K08.html (site discontinued).

White, Michele. 2006. *The Body and the Screen: Theories of Internet Spectatorship*. Cambridge, MA: MIT Press.

Whitlock, Robin. 2017. "Eric Lundgren's 'Phoenix' Electric Car, Built from Waste, Achieves Guiness Book of World Records Title." *Renewable Energy Magazine*, October 20, 2017. https://www.renewableenergymagazine.com/electric_hybrid_vehicles/eric-lundgrena-s-a-phoenixa--electric-20171020.

Wideman, G. 1982. "This Computer Will Do Anything! Limited Only by Your Imagination." *Electronics Technology International*, January 1982, 90–92.

Williams, Raymond. 1983. *Keywords: A Vocabulary of Culture and Society*. Revised ed. London: Flamingo / Fontana.

Wilson, Katherine. 2017. *Tinkering: Australians Reinvent DIY Culture*. Clayton, Australia: Monash University Publishing.

Wiltshire, Alex. 2015. *Britsoft: An Oral History*. London: Read Only Memory.

Wolff, Arnold. 1982. *Basic Programming Do-It-Yourself: An Introduction to Computer Programming Using the "BASIC" Language*. North Ryde, Australia: CCH.

Young, Tom. 2015. "Researching Creative Microcomputing in Australia." YouTube. https://www.youtube.com/watch?v=hR8-10qjoiI.

Zabrs, Paul. 1985. "At the Terminal: Basics of Computer Communications." *Australian Apple Review* 2 (2): 7–8.

Zemon Davis, Natalie. 2008. "The Quest of Michel de Certeau." *New York Review of Books*, May 15, 2008.

Index

Note: Page numbers followed by *f* refer to figures.

16-bit systems, 107, 115–116
1541 Ultimate II, 164
8502 chip, 140

a2_poet, 168–170, 192n7
Adafruit, 175
Adams, Scott, 6
Adventure International, 6, 106
Adventure Probe group, 64
Adventures of Indiana Jones in Wenceslas Square in Prague on January 16, 1989, The, 77
Advertisements
  for Commodore computers, 9, 31f, 32, 35
  for homebrew games, 98, 101f, 112–113
  for program submissions, 108–109, 110f, 111f
Aesthetic dimension, 20–21, 74–77, 81, 84, 96, 136
Ahl, David, 1, 95
Akalabeth, 181n5
Alberts, Gerard, 183n14
Albrecht, Bob, 36
Alien Carnage, 127
Altair, 49, 186n1

Amateur programmers. *See also* Hobbyists
  as characteristic of homebrew game development, 3
  dismissive accounts of, 4–9, 181n5
  professionalism of, 6–7
Amiga
  game development for, 71, 111, 116, 125, 127
  introduction of, 107, 115
Amiga Basic, 119
Amstrad
  distribution of, 43, 62, 115
  game development for, 114–116, 188n4, 189n6
Anachronistic game development. *See* Retro game development
Ant Attack, 158–159
Anthropy, Anna, 4, 148, 181n2
Antic, 164, 165f
"Antic" podcast, 183n13
Apple computers
  Apple II, 27, 66, 121, 189n6
  BASIC programming guides for, 36
  best-selling software for, 27–30, 28–29f
  introduction of, 2
  Macintosh, 49, 107, 189n6
  software developed for, 108
Apple II project, 166–167

Apple II X project, 167
Applesauce, 164, 167
Applied Technology, 40, 132, 136, 138. *See also* Microbee
Applix 1616, 132
Appropriation, 17, 143–144. *See also* Clones and reimplementations
Arcade games
  clones and reimplementations of, 89–95
  in computer history, 8
  prevalence in everyday culture, 60–61, 88–89, 172
Archival sources, 15, 19–20
ArComPro, 45–46
Armstrong, Simon, 35–36, 66, 82, 189n8
Arrow, Selwyn, 82, 132–135
Artifacts, digital cultural, 176–180
Art of Coding campaign, 192n5
"Art of making do," 21–22, 51
Art Software, 126
Asteroids, 157
Astounding Arcade Games for the John Sands Sega (Love and Hancock), 92
Atari microcomputers, 34
  Atari STs, 107
  BASIC programming guides for, 36–39
  best-selling software for, 27–30, 28–29f
  introduction of, 1
  market share of, 7–8
At Home with Computers (Lally), 12
Attack of the Mutant Camels, 188n3
Auction Lots, 45
Audience. *See* Users
Audience studies, 10, 17–18, 22, 74, 79, 146, 173
Ault, Roz, 37, 38
"Aura," 91
Australian Apple Review, 41, 118, 124
Australian Bureau of Statistics (ABS), 25
Australian Commodore and Amiga Review, 118–119, 122

Australian Commodore Review, 32, 41, 45, 119
Australian Home Computer GEM, 118
Australian Microcomputer Magazine, 28–29f, 34
Australian Research Council, 19
AWA, 114

Baer, Ralph, 13
Bar File, 45
Barnes, John G. P., 185n11
BASIC (Beginners All-purpose Symbolic Instruction Code), 71, 156, 186n19
  criticism of, 185n11
  "how-to" guides for, 36–39, 59, 69
  QuickBASIC, 117
BASIC Computer Games (Ahl), 1, 95
BASIC Programming for Kids (Ault), 37
Battlestar Galactica, 86
BBC
  The Computer Programme, 9
  microcomputers, 9, 36, 62, 189n6
Beals, Fiona, 82
  formal school computer classes, 66–67
  games developed by, 77
  gendered access to computers, 61–64
  learning by doing approach of, 68–69
  motivations of, 152
  self-taught programming, 38
Beam Software, 91, 172, 189n6, 193n2
"Bedroom cultures" of coding, 20, 70, 148
Beef Stud File, 45
Bejeweled, 149
Benjamin, Walter, 19, 91, 161, 177
Berlin Computerspiele Museum, 14, 130, 162
Better World studios, 91–92
Bewildered Parent's Guide to Computer Programming, The (Micchia), 38
Bible Baseball, 166
Big Fish games, 149
Birss, Neil, 30, 44

Index

Bits and Bytes, 30, 41, 108–109, 135
Blitz Research, 95
Bogost, Ian, 176
Booth, Paul, 145
Borland, John, 4–9, 181n5
Boulder Dash, 124
Bowden, Robert, 45
Bradfield, Andrew, 164
Breen, Neil, 132–133
Brian, Robert B., 45
Brin, David, 80
Brisbane PC1500 Bit Fiddlers Club, 133
Broderbund Software advertisement, 111f
Brügger, Niels, 191n11
Buchsbaum, Walter, 133
Budgeting software, 26, 184n4
Builders, users as. *See* Hardware hacking; User practices and consumption
"Build your own" computers, 129–135
Bunyip Adventure, 86
Burra Charter, 177
Bushnell, Nolan, 13
Business computers, ambivalence about, 30–33
Butterfield, Jim, 35
Byte, 132, 182n7

Camelli, 188n3
Campbell-Kelly, Martin, 5, 34
Canberra Canberra, 86
Capitalizing, production without, 107–109, 189n10
Capture of Speech and Other Political Writings, The (de Certeau), 55
Cass, Stephen, 115
Catalan re-patching of *Summer Games II*, 169
Center/periphery relationship, in game histories, 14–17, 172, 182n11, 183n14
Champness, Rodney, 190n4

Cheese & the Worms, The (Ginzburg), 182n12
Chilly Willy, 68, 89–91, 95
Chroma interface, 158
"Chronicle era," 13
Cifaldi, Frank, 164
Circuit cookbooks, 136–142
City Lander, 42, 95, 109
C language, 119
Clarkson, Vaughan, 82
   game sales and distribution, 40, 107, 116
   games developed by, 93–95
   influences on, 86
   professional career, 189n8
Cleese, John, 9
Clones and reimplementations, 89–95. *See also* Demakes
   "aura" in, 91
   Chilly Willy, 68, 89–91, 95
   encouragement and acceptance of, 91–92
   Gridfire, 93
   Hoards of the Deep Realm, 93–95
   regional differences in, 188n3
Cockroach Graphics Utility, 119
CoCo. *See* Tandy Color Computer (CoCo)
Coding education, 35–39, 80, 189n8
Collections
   in cultural institutions, 163–164
   unofficial archives, 164–167
Combined Microcomputer Users Group, 131
Commodore. *See also* Amiga; Commodore 64
   advertisements for, 9, 31f, 32, 35
   BASIC programming guides for, 36
   best-selling software for, 184n5
   Commodore 128, 119
   introduction of, 1
   market share of, 7
   PET 2001, 121

Commodore 64
  advertisements for, 9, 35
  BASIC programming guides for, 36
  best-selling software for, 27–30, 28–29f
  business use of, 32
  game development for, 95, 114, 184n5, 189n6
  home use of, 7, 62, 64
  introduction of, 121
Commodore 128 Troubleshooting and Repair (Heilborn), 140
Commodore User, 120
Communications revolution, 118
Compaq, 9
Compu-B, 46
Compucolour, 108
Compute! 9
Computer Assisted Instruction (CAI), 46
Computer Chronicles, The (TV series), 9
Computer Classics, 114
Computer Gaming World, 9
Computer Input, 41, 42, 109
Computer Programme, The (TV series), 9
Computers and Video Games, 41, 120
Computer Seminars, 185n10
"Computing on the Cheap" (Lee), 32
Conley, Tom, 56
Console Living Room, 166
Consoles, home-built, 133–135
Consumer Expenditure Survey (CES), 25
Consumers. *See* Users
Consumption. *See* User practices and consumption
Contemporary retro homebrew. *See* Retro game development
Cooking analogy, 67–69, 191n1
Copy protection, cracking of, 167
Copyright, 86, 89–91, 167, 190n12, 192n4
Cosmic Bomber, 97
Costs, microcomputer, 25–26
Covid-19 pandemic, 175, 179
Cracking, 121, 167, 183n15

Creativity, as motivation for homebrew practice, 84–85
Crooks, Paul, 30
Crossfire, 93
Crossy Road, 150–151, 158, 162, 189n5
CroZXy Road, 158–163
Cultural and fan studies, 22, 78–80, 145, 186n2, 191n11
Cultural institution collections, 163–164
Cultural reception of computers. *See* Discourses of utility
Culture. *See also* Cultural and fan studies
  creation of, 51, 55–56
  digital cultural heritage, 176–180
  elitist conception of, 55–56
  mass versus ordinary, 20–21, 51, 55–56
  popular, 86–87
Curiosity, as motivation for homebrew practice, 83–84
Czechoslovakia, game history in, 14

Daily Telegraph, 112
Dambusters, 124
Dare, The, 85
Davidson, Michael, 186n16
Davis, Natalie Zemon, 50
"Dead media," 22, 151
Deane, John, 39
Deane, Judy, 39
de Certeau, Michel, 19, 49–50. *See also Practice of Everyday Life, The* (de Certeau)
  on "art of making do," 21–22, 51
  Giard and, 50–56, 187n8
  on heterodoxy, 16
  on mass versus ordinary culture, 20–21, 51–52, 55–56
  on multidimensionality of everyday practice, 74–77, 83
  on "ordinary man," 179

on *poiesis* and making, 11, 51, 70, 136, 182n8
on "production without capitalizing," 107, 189n10
on productivist rationality, 77
on "ruses of consumers," 60
on scriptural economy, 52, 144
theory of consumption, 11–12, 50–56, 74–76, 142, 172, 183n15
Decline theses, 21, 80, 82, 119–124, 152, 190n16
Defender, 61
Degiovani, Renato, 41
Demakes, 189n5. *See also* Clones and reimplementations
contemporary development of, 22, 151
copyright and, 192n4
CroZXy Road, 158–163
definition of, 155
Gate Crasher, 154–156
Demoscene ethic, 154, 192n5
Denis through the Drinking Glass, 77
De Simone, Giuseppe, 116, 190n13
Desktop publishing, 119
Dick Smith Electronics
advertisements by, 108–109
Dick Smith Colour Computer, 109
System 80 microcomputer, 1, 36, 108
user groups for, 44
VZ computer, 44
Digital cultural heritage, 176–180
Digital Folklore (Lialina and Espenschied), 10
Digital media histories, 79–80
Digital rights management (DRM), 175
Dinky Kong, 71, 108
Disciplinary stakes, 77–80
Discourses of utility, 20
claimed uses, 25–30
computer magazines, 41–44, 185n15
experimentation and new uses, 39–40, 185n13

gaming, 33–35
home user perceptions, 30–33, 184n6
market penetration and, 25, 27, 184n3
nongame software, 45–47
programming, 35–39
purchase versus actual use, 27–30
reception shaped by, 23–24, 160
serious software users, 30–33
technology in search of a use, 33
user groups, 44–45
user versus nonuser perceptions, 24–25
Discretionary purchases, microcomputers as, 25–26
Disk-based storage systems, 126
Distribution, 6–7
cottage industries for, 109–115
hobbyist mercantilism and, 107–109
production without capitalizing, 107–109, 189n10
shadow economies for, 6–7
Domestic location, of homebrew practice, 3
Donkey Kong, 61, 188n4
Donovan, Tristan, 13, 148
Donut Dilemma, 21
design of, 102–106
inspiration for, 86, 98–99
marketing and advertisement, 98, 101f, 106–107
packaging, 105f, 106
success of, 94, 107
Doom, 144, 156, 183n15
DosBox, 166
DOS operating system, 2
DotSoft, 114
Douglas, Susan J., 175
Doyle, Shayne, 135
Dragon 32, 189n6
Dr Dobbs Journal, 9, 182n7
Dreamcards, 46
DREAM computer, 132
Dungeons beneath Cairo, 109
Dunki, Quinn, 170

DVATS, 117
Dynamic Software, 108

Ecology of games, 87–89
EDUC-8 computer, 132
Educational software, 46, 166, 189n6
Edwards, Benj, 95
Egri-Nagy, Attila, 152
Electric Dreams (Friedman), 49
Electronic Concepts Pty Ltd, 34
Electronic Games (Buchsbaum and Mauro), 133
Electronics Australia, 41, 132
Electronics tinkering. *See* Hardware hacking
Electronics Today International, 41, 132
Elevator Action, 61
Emu Joust, 85
Emulator packages, 166
Ensmenger, Nathan, 67–68
Epps, Garry, 108
Espenschied, Dragan, 10, 191n11
E.T., 13
Ethical dimension, 20–21, 74–77, 81, 84, 96, 136
Ethnography, 10–12, 53, 56, 79, 124
ETI (*Electronics Today International*), 137
Exceptionalism, 16
Exidy Sorcerer, 36, 60, 121, 133, 189n7
Experimental ethic, 3
Experimentation, 21, 39–40, 68–69, 185n13
Experimenter (Series 2 model), 138, 139f

Fackerell, Michael, 108
Fan studies, 19, 21–22, 142–146, 186n3, 191n11
Fantastic Game Books, 37
Farrell, Andrew, 185n7
Fassone, Riccardo, 188n3
Fausett, Sean, 189n6
Female coders, challenges faced by, 61–66

Ferrier, Marie, 183n16
Finnish Museum of Games, 14
Fiske, John, 21–22, 51, 75, 187n4
Fixes, 168–170
Flappy Bird, 150
Flight of the Amazon Queen, 127
Flight Simulator, 27
Foord, Martin, 108
Ford, Lindsay R., 46
For Ectomorphs Only, 45
Foucault, Michel, 52
4am (cracker), 166–170
France, Sharon, 40
Friedman, Ted, 49, 186n1
Frogger, 91, 150–151, 162, 189n5
From Bedrooms to Billions (documentary), 9, 149
Fun, as motivation for homebrew practice, 83–84, 152
Fun Division, 96, 98. *See also* Marentes, Nickolas

Galloway, Patricia, 142
Game consoles, home-built, 133–135
Game development. *See* Homebrew game development; Retro game development
Game ecology, 21, 87–89
Game history, 3, 13–15, 171–172.
   See also Discourses of utility; Game preservation; Homebrew game development
   center-periphery relationship in, 13–17, 172, 182n11, 183n14
   chronicle era of, 13
   continuities and discontinuities with modern practices, 171–173
   "great men," focus on, 13–14, 16, 78
   heterodox approach to, 15–18, 182n12
   as hidden history, 12–13
   interdisciplinary approach to, 18–19
   locality and difference in, 14–15, 172

Index

marginalization of microcomputing in, 4–9, 77–78, 142–145
methods and sources for, 18–22
micro era, 2
microhistorical approach to, 14–15, 182n10
player studies, 10–11
plurality and inclusivity and, 14
standard narrative of, 13
theoretical framework for, 9–12, 74–77, 173
web history, 145, 191n11
Game History and the Local (Swalwell), 182n9
Game Point Software, 109
Game preservation
cultural institution collections, 163–164
Play It Again project, 19, 94, 172
unofficial archives, 164–167
Gameworx, 113
Gaming
computing versus, 121–124
discourse of utility and, 33–35
ecosystem of, 21, 87–89
Garriott, Richard, 6, 181n5, 182n7
Gate Crasher, 154–156
Gazzard, Alison, 8, 188n3
Gendered access, 61–66
Genealogy, 46
General Household Survey (GHS), 25
GeoCities Research Institute, The, 191n11
Gerlach, Gerry, 6
Ghost Busters, 63
"Ghosts in the City" (de Certeau and Giard), 16
Giard, Luce
cooking analogy, 67–69, 191n1
de Certeau and, 50–56, 187n8
on heterodoxy, 16
interview technique of, 183n16
on mass versus ordinary culture, 20–21, 51–52, 55–56

on multidimensionality of everyday practice, 74–77, 83, 136
on satisfaction of making, 70, 73
theory of consumption, 11, 52, 53–55, 142, 172
Giles, Richard, 133–135
Ginzburg, Carlo, 16, 182n12
Girls, coding participation by, 61–66
Gladiator, The, 98
Gleaning, 17
Gloom, 155
Goggin, Gerard, 182n11
"Got Papers?" archive, 192n5
Grandstand Electronics, 41–42, 109, 115, 116, 186n16
Great International Paper Airplane Construction Kit, The, 124
"Great men," focus on, 13–14, 16, 78
Gridfire, 93
Grotnik Software, 86
Groups, user, 44–45, 131, 186n16
Gunning, Tom, 160
Gutman, Dan, 182n7

Hacking. *See* Hardware hacking
Hacking Europe (Alberts and Oldenziel), 183n14
Hadrup, Michael J., 44, 116
Haigh, Thomas, 7, 8, 173
Halegoua, Germaine, 182n11
Hall, Matthew, 82, 125, 148–150, 189n8
Halloween Harry, 68, 84
Ham radio, 135–136, 190nn3–4
Hansen, Miriam, 177
Harboro, 86
Harbour, 109
Hardware hacking, 18, 136–142
archival traces of, 130
"art of making do," 21–22, 51
circuit cookbooks for, 136–142
contemporary practices informed by, 146
ethics of, 133, 137–138

Hardware hacking (cont.)
  ham radio, 135–136, 190nn3–4
  historical marginalization of micro users in, 142–145
  hobbyist interest in, 121
  home-built hardware, 129–135
  new research directions for, 175–176
  prevalence of, 190n1
  recovering role of early micro users in, 145–146
  scarcity of computer components and, 130–131, 190n2
  as "serious leisure," 135–136
Hari, Kristen, 190n3
Haring, Kristen, 190n3
Hartnell, Tim, 83, 188n1
Heilborn, John, 140, 191n5
Heritagization process, 16
Heterodoxy, 15–18, 182n12
Hilbert, Ernest, 161
Hill, Matt, 145
Hipster Whale, 150, 159–160. *See also* Crossy Road; CroZXy Road
Hoards of the Deep Realm, 93–95
Hobbit, The, 124, 189n6
Hobby Computer Club (HCC), 120–121
Hobbyists. *See also* Hardware hacking; Users
  in discourse of utility, 24–25
  experimentation by, 40
  types of, 135–136, 190n15
Hodgson, Dean, 46
Hoess, Rudi, 26, 34, 35
Homebrew, definition of, 2–4, 70
Homebrew Computer Club, 3, 181n1, 191n5, 191n7
Homebrew game development. *See also* Legacy of homebrew gaming; Retro game development; User practices and consumption; *individual games*
  characteristics of, 2–4
  clones and reimplementations, 89–95, 188n3
  continuities and discontinuities with modern practices, 171–173
  culture of, 70–73, 179–180
  decline theses, 21, 80, 82, 119–124, 152, 190n16
  demakes, 22, 151, 154–156, 158–163, 189n5, 192n4
  evaluative criteria for, 181n3
  experimentation in, 21, 39–40, 68–69, 185n13
  gaming ecosystem for, 21, 87–89
  homemade aesthetic of, 70–73
  improvisation in, 40, 69, 81, 92, 137
  influences on, 85–87
  mainstreaming of personal computing through, 12–13, 34–35, 117–119
  marginalization of, 4–9, 77–78, 181n6, 181nn3–4
  motivations for, 73–74, 83–85
  nostalgia for, 13, 153, 160–161
  ordinary culture in, 20–21, 173–174
  polemic/aesthetic/ethical dimensions of, 20–21, 74–77, 81, 84, 96, 136
  popularity from mid-1980s on, 115–116, 124–126
  ported games, 91–92, 106, 189n6
  quality judgements about, 4–9, 181nn3–4
  significance of, 4, 6, 78–80, 181n2
  teamwork in, 126–127
Homebrew game distribution, 6–7
  cottage industries for, 109–115
  hobbyist mercantilism and, 107–109
  production without capitalizing, 107–109, 189n10
Home-built hardware, 129–135
  archival traces of, 130
  computers, 132–133
  contemporary practices informed by, 146
  ethics of, 133, 137–138
  game consoles, 133–135
  keyboards, 130–131

Index

prevalence of, 129–130
as "serious leisure," 135–136
Home computers, perceived utility of, 30–31, 184n6
Home console platforms, 8
Home environs
gendered access in, 61–66
micro users as makers in, 57–61
Homemade aesthetic, 70–73
Homogenization, 51
Honeysoft, 40, 93–95, 116
Hopper, Grace, 68
Horace Goes Skiing, 91
Hot Copter, 164, 165f
Household Use of Information Technology (ABS), 25
"How-to" programming guides, 36–39, 59, 69
How Users Matter (Oudshoorn and Pinch), 9
Huhtamo, Erkki, 13, 17, 88

IBM computers, 2, 116
best-selling software for, 27–30, 28–29f
business market for, 33
IBM 286 PCs, 107
IFixit, 175
Image Link, 117
Imagineering, 27, 34
Impact! 157
Improvisation, 40, 69, 81, 92, 137
Inclusivity, 14–15
Incorporation/resistance paradigm, 75
Indie/informal game development, homebrew as inspiration for, 4, 148–151, 181n2
Informants, 82
Innervation, 177, 193n3
Institute of Backyard Studies, 175
Intangible Cultural Heritage (UNESCO), 192n5
Interactive Binary Illusions, 127
Interdisciplinarity, 18–19

Interface Publications, 83
International Council on Monuments and Sites (ICOMOS), 177
Internet Arcade, 166
Internet Archive, 118, 164, 166, 167
Internet history, 145, 191n11
Interviews
informants, 82
research methodology, 19–20, 183n16
Ironmonger, D. S., 25, 27

Japan, as perceived center of game development, 13, 14, 172
Jenkins, Henry, 22, 52, 75, 143, 145, 191n10
Jenner, Caitlyn, 170
Jewels of Sancara Island, 125
Jobs, Steve, 3
Johnson, Steven, 144
Jones, Stephen, 193n2
Jones, Steven, 143, 144

Kenyon, Philip, 41–42
Keogh, Brendan, 150
Kerr, Andrew, 109
Keyboards, home-built, 130–131
Kilobaud, 9
King, Brad, 4–9, 181n5
Kirkpatrick, Graeme, 8, 21, 82, 119–124, 181n3, 182n7
Kitch, Bob, 44
Kit computers, 132–133
Kiwi Computer, 71
Kong Tin, Harvey, 164
Kowalski, John, 155–156
Kryoflux, 164

Lab Link, 117
Lally, Elaine, 12
Lange, Andreas, 130
Laser Hawk, 164
Laughton, Alan, 94
Laws, John, 9, 35

Lean, Tom, 8, 123, 181n4, 190n15
Lee, Frank, 32
Legacy of homebrew gaming, 21–22
  anachronistic and retro coding practices, 151–163
  cultural institution collections, 163–164
  definition of, 147–148
  8-bit aesthetic, 192n2
  impact of professionalization on, 191n1
  inspiration for contemporary game development, 4, 148–151, 181n2
  patches and fixes, 168–170
  unofficial archives, 164–167
Legitimacy, ordinary culture and, 77, 163, 172, 174, 192n5
Leibovitz, Annie, 170
Letraset, 71, 98
Levi, Giovanni, 15
Lexaloffle, 187n12
LGBTQ issues, 14
Lialina, Olia, 10, 191n11
Lindley, Richard, 98
Lindsay, Christina, 24
Lindsay, Eric, 27, 33, 117, 136, 137
L'invention du quotidien. See Practice of Everyday Life, The (de Certeau)
Lloyd-Smith, C. W., 25, 27
Lobato, Ramon, 150
Local environs
  gendered access in, 61–66
  micro users as makers in, 57–61
Locality, 14–15, 86, 172, 182n9
Lode Runner, 93. *See also Hoards of the Deep Realm*
Longhurst, Brian, 74
L'ordinaire de la communication (de Certeau and Giard), 54
"Losing Meanings" (Veraart), 120
Lowood, Henry, 176
Lumiere Brothers, 160
Lunar Lander, 95

"Machine Code Made Easy," 39
Macinitizer, 125
Macintosh, 49, 107, 189n6
Magazines, 41–44, 118–119, 185n15. *See also individual magazines*
Magnetic tape technology, 125
Mainstreaming of personal computing, 12–13, 34–35, 117–119
Maker Faires, 51, 186n3
Makers, micro users as. *See* User practices and consumption
MAME (Multiple Arcade Machine Emulator), 166
Marentes, Nickolas, 82, 189n8. *See also Donut Dilemma*
  on Amiga, 116
  commercial success of, 108, 109, 125–126
  Cosmic Bomber, 97
  Donut Dilemma, 21, 86, 94, 96–106
  education, 67, 71
  Gate Crasher, 154–156
  The Gladiator, 98
  marketing and market research, 97–98, 106–107
  Moon Scout, 98
  motivations of, 96–97, 111–112, 122
  Neutroid, 96, 98, 112, 189n9
  Pac-Man Tribute, 96, 154
  *Pop*Star Pilot*, 153–154
  professional approach of, 97–106
  Rupert Rythym, 107
  Space Intruders, 107
  Stellar Odyssey, 98
  working method, 98–106
Mark-8 computer, 132
Marketing, 106–107
  advertisements, 98, 112–113
  market research, 97–98
  sales brochures, 99–100f
Market penetration, 25, 27, 184n3
Mass culture, 51
Mass market users, 123, 190n15

Index

Maths Invaders, 46
Mauro, Robert, 133
Maynard, F. K., 43
Mayol, Pierre, 11, 20, 50, 52–56, 172, 183n16
McCracken, Harry, 95
McGinley, Jim, 150–151
McKechnie, Cameron, 71, 126
Meaning of Video Games, The (Jones), 143
Media archaeology, 4, 17–18, 174
Media studies, 146
  contributions of homebrew to, 78–80
  media ethnography, 53
Mendham, Trevor, 35
Menocchio, 182n12
Mercantilism, 107–109
Merlin, 46
MESS, 166
Micchia, Shane, 38
Microbee
  BASIC programming guides for, 36
  computer kits, 132, 136–137
  decline of, 116, 190n13
  Exidy Sorcerer compared to, 189n7
  Experimenter (Series 2 model), 138, 139f
  game development for, 40, 68, 93–95, 108–109, 125
  introduction of, 1
  transfer of programs between, 135
Microbee Hacker's Handbook, 138, 140f, 141f
Microcomputer Enthusiasts Group (MEG), 135
Microcomputers, public discourses about. *See* Discourses of utility
Microcomputing decline theses, 21, 80, 82, 119–124, 152, 190n16
Micro era, 2
Microforte, 172, 193n2
Microhistorical studies, 14–15, 182n10
Micro Sistemas, 41

Microsoft
  DOS operating system, 2
  home console platforms, 8
Micro users. *See* Users
Microwave Oven Troubleshooting and Repair (Heilborn), 140
MikroBitti, 188n3
Millard, Dorothy, 82
  games developed by, 85–86, 125
  motivations of, 64–66, 83
Millington, Glenys, 43
Miner Man, 157
Minitel, 54
Mod-chips, 18
Modding, 17, 176, 183n15, 191n6
Monro, Don, 37
Montford, Nick, 176
Moon Patrol, 61
Moon Scout, 98
More TRS-80 BASIC (Albrecht, Inman, and Zamora), 36
Morris, John Keoni, 167
Mosaic, 145
Mozzie Zapper, 85
MSX, 189n6
Multidimensionality of everyday practice, 74–77, 153
Mystery of Munroe Manor, 115
Mytek Computing, 85, 109, 110f, 111f

"Naive Users," 10
National heritage, politics of, 16
Neil, Katharine, 25–26, 36, 61–64, 82, 189n8
Nelson, Andrew, 188n1
Neutroid, 96, 98, 112, 189n9
New research directions, 22
  contemporary vernacular digitality, 174–176
  continuities and discontinuities with homebrew gaming, 171–173
  digital cultural heritage, 176–180
  ordinary culture and, 173–174

Nintendo, 8, 115, 188n4, 192n6
Noir Shapes, 157–158
Nongame software, 45–47
Nonusers, perceptions of microcomputer utility, 46–47
Nostalgia, 13, 161
NSW Wireless Institute Centre, 135–136

O'Hara, Rob, 9, 183n13
Oldenziel, Ruth, 183n14
Olympic Decathlon fix, 170
Olympic Gold, 46
Online: The Microbee Owner's Journal, 27, 41, 45, 138
Operativity/operations, 187n11
Oral history interviews. *See* Interviews
Ordinary culture, 20–21, 51–52, 55–56, 173–174
Ordinary practice, microcomputing as, 55–56
  gendered access, 61–66
  in home and local environs, 57–61
  at school, 66–67
Origin Systems, 6
Oudshoorn, Nelly, 9

Pacific Computer Weekly (PCW), 32–33
Packaging, 105f, 106
Pac-Man, 92
Pac-Man 256, 150
Pac-Man for the Vic-20, 188n3
Pac-Man Tribute, 96, 154
Parikka, Jussi, 147
Passfield, John, 82, 94, 189n8, 190n16
  Chilly Willy, 68, 89–91, 95
  commercial success of, 40, 107, 111–112
  Halloween Harry, 84, 127
  motivations of, 76, 84–85
  16-bit system games, 116
  team collaboration, 127
Passport, 167
Patches, 168–170

PBS, *The Computer Chronicles*, 9
PDP-11, 135
Pengo, 89. *See also Chilly Willy*
People's Computer Company, 36
People's History of Computing in the United States, A (Rankin), 174
Period, 45
Peripherals, 137
Peripheries, in game histories, 14–17, 172, 182n11, 183n14
Perry, John, 57, 82
  City Lander, 42, 95, 109
  commercial success of, 107, 109
  Harbour, 109
  learning by doing approach of, 69
Personal computer (PC)
  game development for, 189n6
  introduction of, 2
  mainstreaming of, 12–13, 34–35, 117–119
  microcomputers as predecessor to, 171
Perspective Software, 71
PET 2001, 121
Philipson, Graeme, 24, 33, 40
Phillips P2000, 121
Pinch, Trevor, 9
Planning Tanning, 45
Play It Again project, 19, 94, 172
Pluralization, 52
Poaching, 22, 52, 143–145, 191n10
Poiesis, 51, 182n8
Polemical/political dimensions, 20–21, 74–77, 81, 84, 96, 136
Pong, 59–60
Pony Jamboree, 45
*Pop*Star Pilot*, 153–154
PopCap, 149
Popular culture, influence of, 86–87, 144
Popular Memory Archive. *See* Play It Again project
Porting, 91–92, 106, 189n6
Poseidon Software, 43–44

Index

Poster, Mark, 74
Powell, Gareth, 118–119
"Practical Science of the Singular, A" (de Certeau and Giard), 54–55
Practice of Everyday Life, The (de Certeau), 50–56, 187n5, 187n8. *See also* De Certeau, Michel
Presentism, 4
Printer Troubleshooting and Repair (Heilborn), 140
Production, consumption as form of, 50–56. *See also* User practices and consumption
Productivist rationality, 77
Productivity software, 119
Professionalization, impact of, 191n1
Psychotec, 46

QuickBASIC, 117
Quill, The, 82
Quizmaster, 46

Radio. *See* Ham radio
Radio Electronics, 132
Radio Operator's Logbook, 45
Radio Shack. *See* Tandy Color Computer (CoCo); Tandy Radio Shack-80 (TRS-80)
Rag Doll Kung Fu, 149
Rainbow, The, 109, 185n15
Rankin, Joy, 174
Rare Trades (Thompson), 162
Raspberry Pi, 4, 151
Reagan, 77
"Real Users," 10
Reimplementations. *See* Clones and reimplementations
"Remembering and the Forgetting of Early Digital Games, The" (Swalwell), 160
Remixing, 17
Rendall, Steven, 50
Replay (Donovan), 13

Research methodology. *See also* New research directions
heterodox approach to, 15–18, 182n12
implications of, 22
interdisciplinarity of, 18–19
interviews, 19–20
sources, 18–22
theoretical framework for, 9–12, 74–77, 172–173
Resistance paradigm, 75–77
Retro game development, 22, 151–163, 192n2
Ant Attack, 158–159
CroZXy Road, 158–163
8-bit aesthetic in, 192n2
Gate Crasher, 154–156
homebrew era as inspiration for, 4, 17, 148–151, 181n2
Impact! 157
as living culture, 170, 179
motivations for, 152–153
Noir Shapes, 157–158
Pac-Man Tribute, 154
*Pop*Star Pilot*, 153–154
Stranded, 156–157
Reynolds, Darryll, 30, 82, 189n8
commercial success of, 108, 113–115
games developed by, 64, 85, 115, 116
influences on, 86–87
nongame software developed by, 69
Richardson, Ric, 32
Right to repair movement, 175
Rise of the Videogame Zinesters (Anthropy), 4
Roe, James, 46
Rowe, Jamieson, 132, 135, 190n4
Rowlands, Grant, 35
Rupert Rythym, 107

Salon, 80
Sands, John, 44

Satisfaction, as motivation for homebrew practice, 73–74
Savetz, Kevin, 9, 183n13
Scandella, Domenico, 182n12
SCART socket, 158
Schellens, John, 46
Schlombs, Corinna, 17
Schmitt, John, 25
Schools, microcomputing in, 66–67
Scott, Jason, 166–167
Scriptural economy, 21, 52, 144
Search for King Solomon's Mines, The, 64, 65f
Secret of Bastow Manor, 115
Sega
 home console platforms, 8
 Master System, 44
 Mega Drive, 115
 SC3000, 41, 95, 109, 126, 186n16
Sega Computer, 41–44, 108, 116, 185n15
Sega Programming Manual (Brian), 45
Sega Users Club, 186n16
Selecta-Game, 133
Seremetakis, C. Nadia, 4, 161
Series 2 model. *See* Experimenter (Series 2 model)
"Serious leisure," 135–136
"Serious software people," 30–33
Severn Software, 115
Sharemarket, 46
Shatner, William, 9
Shklovsky, Victor, 160
Showjump, 46
Sibly, Mark, 82, 189n8
 games developed by, 71, 95, 108, 116, 126–127
 learning by doing approach of, 35–36, 66
 team collaboration, 126–127
Sierra Online, 6
"Silent" cracks, 167
Silicon Chip, 190n4
Silverstone, Roger, 75

Sinclair, 1. *See also* ZX Spectrum
 ZX80, 121, 122
 ZX81, 59, 156–158, 161
Sirius 7, 126–127
16-bit systems, 107, 115–116
Smith, Bob, 189n8
 copyright and, 192n4
 games developed by, 156–163, 192n3
Smith, Mathew, 148–149
Smith, Rodney, 126
Social aspects of coding, 152
Softalk, 9, 182n7
Softgold, 113
Softime (NZ) Limited, 189n6
Software heritage projects, 22
Sony home console platforms, 8
Sorcerer. *See* Exidy Sorcerer
Sorceror's Apprentice, 71, 126
Soupourmas, F., 25, 27
South, Raumati, 135
Space Adventure Books, 37
Space Intruders, 107
Space Invaders, 13, 92
Spectrum. *See* ZX Spectrum
Spelling, 46
Squash Controller, 46
Stamatiadis, Steve, 127
Stebbins, Robert A., 135–136, 164
Stellar Odyssey, 98
Stephen, Andrew, 57, 59, 82, 189n8
Stephen M. Cabrinety collection (Stanford), 9
Storer, Jim, 95
Stranded, 156–157
Strategic Studies Group, 172
Street, C. A., 185n13
Streeter, Arthur, 82, 189n8
 commercial success of, 77, 108, 112–113, 190n12
 games developed by, 85
 nongame software developed by, 116–117
Street Games. *See* Streeter, Arthur

Index

Stuckey, Helen, 189n6
Summer Games II, 168–170, 178
Sun Herald (Sydney), 37
Super 80, 132
Super Mario Bros, 192n6
Super Nintendo Entertainment System, 115
Supersoft Software, 96, 97. *See also* Marentes, Nickolas
Švelch, Jaroslav, 8
Sydney Morning Herald, 112
Symons, Ross, 82, 189n8
  books authored by, 71, 73, 76, 107, 188n1
  on indie game development, 148, 150
  learning by doing approach of, 57, 59–60
  motivations of, 83
Synchrotech Software, 117
System 80 microcomputer, 1, 36, 108

TAFE (technical education) courses, 71
Taito, 132
Tandy Color Computer (CoCo)
  BASIC programming guides for, 36–39
  educational applications for, 46
  game development for, 96, 109, 153, 154, 156, 161
Tandy Radio Shack-80 (TRS-80), 121
  BASIC programming guides for, 36
  game development for, 39–40, 108, 113, 150
  introduction of, 1
  target market of, 24
Teamwork, 126–127
Technological protection mechanisms (TPMs), 18, 175
Telecom, 127
Tester, Ross, 190n4
TestSheepNZ, 92
Texas Instruments, 7, 36–39
Textual poachers, 22, 52, 143–145, 191n10

Textual Poachers (Jenkins), 22, 52, 145
Thatcher, Denis, 77
Theoretical framework, 9–12, 74–77, 172–173
Thermonuclear WarGames, 87
Thomson, Mark, 162, 175
Tinn, Honghong, 142
Tomasik, Timothy J., 53, 187n8
Top Half, 42
TRS-80. *See* Tandy Radio Shack-80 (TRS-80)
Tumblr, 178–179
TVNZ, *Top Half*, 42
Typing tutor software, 30

U-Bend, 158
UNESCO Intangible Cultural Heritage list, 192n5
United Kingdom
  General Household Survey (GHS), 25
  as perceived center of game development, 9, 14, 182n7
United States
  Consumer Expenditure Survey (CES), 25
  as perceived center of game development, 13–14, 17, 172, 183n14
Unofficial archives, 164–167
Usborne programming guides, 38
User groups, 44–45, 131, 186n16
User practices and consumption, 11–12, 49–50. *See also* Hardware hacking; Homebrew game development
  cooking analogy, 67–69
  disciplinary stakes of, 77–80
  experimentation in, 68–69
  gendered access in, 61–66
  heterodox approach to, 17–18, 183n15
  in home and local environs, 57–61
  homebrew programming, 70–73
  improvisation in, 40, 69, 81, 92, 137
  incorporation/resistance paradigm of, 75–77

User practices and consumption (cont.)
  new research directions for, 175–176
  personal satisfaction gained from, 73–74
  polemic/aesthetic/ethical dimensions of, 20–21, 74–77, 81, 84, 96, 136
  at school, 66–67
  theory of consumption and, 11–12, 50–56, 74–76, 142, 172, 183n15
Users. *See also* Hobbyists; User practices and consumption
  definition of, 10
  home, 30–33, 184n6
  perceptions of microcomputer utility, 24–25, 46–47
  "serious software people," 30–33
  terminology and historicity of, 10–12
Utility, discourses of. *See* Discourses of utility

VCR Troubleshooting and Repair Guide (Heilborn), 140
Veitch, Jeff, 86
Veraart, Frank, 21, 82, 119–124. *See also* Decline theses
Vernacular digitality, 23
  contemporary characteristics of, 174–180
  digital cultural heritage and, 176–180
  microcomputers and, 12–13
Viatel, 118
Vic-20
  advertisements for, 9
  BASIC programming guides for, 36
  best-selling software for, 27–30, 28–29f
  game development for, 87, 108, 112–113
Video Games (Buckwalter), 133
Vignau, Antoine, 167
Vintage games. *See* Retro game development

Visual Basic, 4
VZ computer, 44

Wade, Alex, 20
Wadsworth, Jonathan, 25
WarGames, 86–87
Wargaming, 14
Warranty Recorder, 46
Wasiak, Patryk, 190n2
Ways of operating, 187n11
Web 2.0, 17, 143
Web history, 145, 191n11
Weight Recorder, 46
Wellington Sega User Club, 45
West Coast Computer Faires, 9, 51
Westgate, Nick, 189n6
Wheeler, Aaron, 186n16
White, John, 60–61, 82, 94, 187n12, 189n8
White, Joseph, 60–61, 187n12
White, Michele, 191n12
Wideman, G., 26
Wilkinson, Jim, 135
Williams, Ken, 6, 182n7
Williams, Raymond, 19, 69
Williams, Roberta, 6, 182n7
Williams, Ross, 86
Windows 95, 145
Wolfenstein 3D, 156
Women, coding participation by, 61–66
"Work of Art in the Age of Mechanical Reproduction, The" (Benjamin), 91
Workplace computers, 30–33
Wozniak, Steve, 3

Young, Denis, 135
Young'ns, 86
Young Ones, The, 86
Your Computer, 32, 41, 98, 130, 132

Index

Z80 machine language, 119
Zabrs, Paul, 118
Zork I, 27
Zuppicich, Blair, 126
ZX80, 121, 122
ZX81, 59, 156–158, 161
ZXagon, 192n4
ZX Spectrum, 63, 92, 156, 185n13
  game development for, 92, 156–159, 189n6, 192n3
  "how-to" programming guides for, 36, 39
  learning by doing approach to, 59, 63
Zzap!64, 41